Algebra II

by Carolyn Wheater

ALPHA
A member of Penguin Random House LLC

Publisher: Mike Sanders
Associate Publisher: Billy Fields
Executive Acquisitions Editor: Lori Hand
Development Editor: Kayla Dugger
Cover Designer: Laura Merriman
Book Designer: William Thomas
Production Editor: Jana M. Stefanciosa
Compositor: Brian Massey
Proofreader: Laura Caddell
Indexer: Brad Herriman

First American Edition, 2015
Published in the United States by DK Publishing
6081 E. 82nd Street, Indianapolis, Indiana 46250

Copyright © 2015 Dorling Kindersley Limited
A Penguin Random House Company
15 16 17 18   10 9 8 7 6 5 4 3 2 1
001–285170–November/2015

Published in the United States by Dorling Kindersley Limited.

IDIOT'S GUIDES and Design are trademarks of Penguin Random House LLC

ISBN: 978-1-61564-864-1
Library of Congress Catalog Card Number: 2015937087

DK books are available at special discounts when purchased in bulk for sales promotions, premiums, fundraising, or educational use. For details, contact: DK Publishing Special Markets, 345 Hudson Street, New York, New York 10014 or SpecialSales@dk.com.

Printed and bound in the United States

idiotsguides.com

Contents

Introduction

Sequels, as anyone who has gone to the movies knows, are a tricky thing. A successful film may lead to a whole collection of follow-ups, each better than the one before. On the other hand, a second film, based on a fabulous previous hit, may leave you wondering what in the world the filmmakers were thinking. You never know how things will turn out.

Likewise, I have no way to know if your first algebra course was a hit or a flop. I'll assume that if you're tackling algebra II, you had some reasonable success the first time around, but I don't know if you loved it, if every moment was a struggle, or if it was something in between. So trying to craft a sequel you'll enjoy is a tricky matter.

I know you can handle algebra II. A teacher's job is to make that happen, to find the ways to present material, to explain, and to guide you until you have a firm understanding. The thing that's difficult, from a teacher's viewpoint, is that any sequel course is built on assumptions about what you learned in the first course. What I think you should have learned in algebra I and what you actually learned (and what you remember) may be very different. While you're unfortunately not able to tell me, I'm going to try to make sure that you have the essential background, and that my explanations are clear and straightforward.

It's time to get down to the work this book was written for: algebra II. I hope this will be clear and helpful to you, and that you will enjoy your adventures in algebra.

How This Book Is Organized

This book is presented in six parts. Each part contains a chapter or chapters that have some common themes. Here's a roadmap of where you're about to go.

Part 1, Foundations, is where I try to make sure that, when it comes to your previous experience in algebra, you and I are on the same page. In the first chapter, you'll review what I consider the essentials, and then, in the second chapter, I'll present you with functions. Functions may or may not have come up in algebra I. If they did, it was probably just an introduction and not a thorough look at them. Functions are crucial for algebra II and beyond, so this will give you a deeper look.

In **Part 2, Linear Relationships,** you'll look at linear functions and linear inequalities. You'll also look at two types of functions built from linear functions: absolute value functions and piecewise functions. You'll investigate systems of linear equations, expand them to three-variable systems, and explore matrices as a device to solve systems easily.

Algebra II is all about extending ideas from algebra I, so in **Part 3, Quadratic Relationships,** you'll look at techniques of solving quadratic equations that may be new to you. You'll also investigate techniques for quick graphing, and meet some quadratics that are not functions but have interesting graphs.

Part 4, Polynomial and Rational Functions, is the primary extension section of algebra II. If you ever worked with a quadratic function and wondered what would happen if there were terms of a degree higher than 2, this is your part. You'll learn techniques for solving higher-degree equations, you'll investigate their graphs, and then you'll build another family of functions by making quotients of polynomials.

Part 5, Calculating and Counting, you'll look at two chapters that introduce new ideas to solve old problems and give you tools to look at phenomena in the world around you. Exponents and logarithms let you explore unbounded growth or decay, counting theory makes very large numbers manageable, and probability helps you understand your chances.

Part 6, Extra Practice, is just what it sounds like. Reading about math is good, but learning math requires doing. You don't know if you have the skills or not until you try to use them, and you can't solidify them without repetition. You'll find some problems along the way, but this chapter provides the extra you may need.

Extras

As you work your way through this book, you'll see some items set off in ways meant to catch your attention. They're meant to help you on the journey. Here's a summary of what you'll see.

 CHECKPOINT

This is where you'll find some problems. They will let you stop from time to time and take stock of what you know. Checkpoints will ask you to answer a few questions to see if you're ready to move on. You'll find the answers for these checkpoint questions in Appendix C.

 **DEFINITION**

Clear definitions are important to understanding concepts in algebra II. The definition sidebars throughout this book will show you critical words and phrases you'll want to know and use. These and others will appear in the glossary as well.

ALGEBRA TRAP

Algebra traps are a little like speed traps on the highway. They are meant to serve as a warning, and prompt you to think and act carefully. They highlight common errors and areas of difficulty.

ALGEBRA TIP

The flip side of an algebra trap is the algebra tip. These may be bits of information that are helpful to know, but many times you don't learn until you've had some experience. Watch for sidebars throughout the book that point out those bits of information to give you the benefit of others' experience.

CALCULATOR CORNER

Once upon a time, an algebra II book wouldn't have needed these—no technology was necessary. Today, however, calculators, computers, websites, and phone apps have changed the way things are done, or provide quick-and-easy options. Tables to look things up are a thing of the past, and today's graphing calculators can produce, in seconds, graphs that once would have taken far longer. These sidebars are meant to point out places where the technology available to you can be helpful.

Acknowledgments

My gratitude goes to Grace Freedson of Grace Freedson's Publishing Network, who connects me to projects, like this one, that are satisfying and challenging and help me to grow as a teacher and as a person. My thanks also go to Lori Hand and Kayla Dugger, for making this project an absolute delight from start to finish, and for making my ramblings about math look good and make sense.

One of the things I tell my students is that it's normal, natural, and valuable to make mistakes. It's how we learn. More accurately, correcting our mistakes is how we learn. We all make mistakes. I certainly do. The technical reviewer's job is to read everything I've written about the math in this book and make sure it's correct and clear. That job also includes checking all the problems and the answers, and finding my mistakes. I am grateful to my technical reviewer, Cody Baker, for finding those mistakes and pointing them out to me. Cody's corrections and advice make this a better book for you.

To all the folks who had a hand in producing this book, including and especially those whose names I don't know, I send my thanks for all the hard work you do and the pride you take in your work.

Special Thanks to the Technical Reviewer

Idiot's Guides: Algebra II was reviewed by an expert who double-checked the accuracy of what's presented here to help us ensure learning algebra II is as easy as it gets. Special thanks are extended to Cody Baker, who is a graduate student working toward his PhD in Applied and Computational Mathematics and Statistics at the University of Notre Dame.

Foundations

Algebra II builds on algebra I—the name makes that clear. So it's natural to begin with a look back at the critical topics from that first course. In Part 1, you review, remember, and sharpen those earlier skills.

You also spend some time getting comfortable with the concept of a function, the analysis of functions, the arithmetic of functions, and the graphs of functions. As you may suspect, functions will be crucial to all of algebra II.

Building Blocks from Algebra I

The name *algebra II* points to where you need to start. Algebra II builds upon and extends ideas learned in a first algebra course. Ideally, you remember everything about algebra I, but for most people, that's not the case, so a bit of review is in order. In this chapter, I give you a look at the essential algebra I ideas you need to know before beginning algebra II.

The Real Numbers

The first thing you need is a sense of the landscape in which you're working. Algebra uses variables to represent numbers, so understanding the system of real numbers and its various subsets is a sensible place to start.

Real numbers: This set is built up from the simplest experience of counting to the collection of all the numbers with which you can work.

Natural numbers **or counting numbers:** These are the numbers children first use to count: 1, 2, 3, and so on. The natural numbers do not include any fractions or decimals and any negative numbers. They don't even include 0.

Whole numbers: If you add 0 to the natural numbers, you create the whole numbers set: {0, 1, 2, 3, 4 …}.

Integers: These are the positive and negative whole numbers and 0: {… −4, −3, −2, −1, 0, 1, 2, 3, 4 …}. The integers include the natural numbers and the whole numbers as subsets.

Rational numbers: These take their name from the word *ratio*. They include all numbers that can be written as ratios or fractions. This includes decimals that terminate or that repeat a pattern, because they can be converted to fractions. Because all the integers can be written as fractions by putting them over a denominator of 1, the integers are a subset of the rationals.

Irrational numbers: They are just what their name says—numbers that aren't rational. These are numbers that can't be expressed as fractions, and although it may not be immediately obvious what numbers those are, you meet them when you work with circles (because π is irrational) and when you work with square roots. The classic proof that irrational numbers exist is to show that there is no fraction equal to $\sqrt{2}$. Decimals that do not terminate and have no repeating pattern cannot be converted to fractions, so they are irrational as well.

 DEFINITION

Natural numbers is the name given to a set of numbers that include 1, 2, 3, and all larger whole numbers. When 0 is added to the natural numbers, the new set is referred to as **whole numbers. Integers** include the positive whole numbers, the negative whole numbers, and 0. **Rational numbers** are all numbers that can be written as a quotient, or ratio, of two integers. **Irrational numbers** include any numbers that cannot be expressed as a ratio of two integers.

The irrationals are not a subset of the rationals, and the rationals are not a subset of the irrationals. They are disjoint sets with nothing in common, but together, they make up the real numbers. All real numbers are rational or irrational—never both.

CHECKPOINT

Place each number in all the sets to which it belongs: natural, integer, rational, irrational, real.

1. −13

2. 14.1

3. $-\frac{10}{3}$

4. 12

5. $\sqrt{17}$

Simplifying Expressions

Once you have numbers, it's natural to want to do things with them. The four basic operations are addition, subtraction, multiplication (with exponents used as shortcuts for repeated multiplication), and division. The key to performing these operations is to get each individual operation done correctly in the proper order.

The *order of operations* is an agreement that you first deal with multiplication and division as you meet them left to right; you then begin at the left again and do addition and subtraction as you encounter them. If exponents (such as 5^3, with 3 being the exponent) appear, deal with them before any of the four basic operations because they represent repeated multiplication. When parentheses or other grouping symbols appear, do the operations inside them first, following the order of exponents—multiplication and division, and then addition and subtraction. If there are grouping symbols within grouping symbols, work from the innermost set to the outer.

 **DEFINITION**

> The **order of operations** is an agreement that the order for simplifying a problem is expressions in parentheses, exponents, multiplication and division as they appear from left to right, and finally addition and subtraction from left to right.

The order of operations is commonly remembered by the mnemonic PEMDAS, or Please Excuse My Dear Aunt Sally, which stands for Parentheses, Exponents, Multiplication and Division, and Addition and Subtraction.

Order of Operations Example

Let's look at an example of how the order of operations works. To simplify $3 - (4 + 2 \cdot 5) + 6^2 \div 3$, first work in the parentheses, doing the multiplication first, and then the addition. Once those are done, the parentheses are no longer needed.

$$3 - (4 + 2 \cdot 5) + 6^2 \div 3$$
$$3 - (4 + 10) + 6^2 \div 3$$
$$3 - 14 + 6^2 \div 3$$

The next step is to deal with the exponent.

$$3 - 14 + 6^2 \div 3 = 3 - 14 + 36 \div 3$$

Look next for multiplication and division, left to right. In this problem, there's no multiplication, but there is a division.

$$3 - 14 + 36 \div 3 = 3 - 14 + 12$$

Start from the left once again and do addition and subtraction as you meet them.

$$3 - 14 + 12 = -11 + 12 = 1$$

The good news is that most expressions don't involve every rule, as this example did. But it gives you a good idea of how you can execute the different operations in a problem in the correct order.

Order of Operations with Variables

In algebra, you're not simply working with numbers; variables always get involved. So let's look at an example that includes variables. Simplifying $-3x\left(x^2 - 4x + 5\right) + \frac{9x^3 + 6x^2 + 12x}{3x}$ will involve several steps.

Start with $-3x\left(x^2 - 4x + 5\right)$. Remember from algebra that you can only add or subtract like terms. There are no like terms in the parentheses, so there's nothing to do here but use the *distributive property* to multiply $(x^2 - 4x + 5)$ by $-3x$.

$$-3x\left(x^2 - 4x + 5\right) + \frac{9x^3 + 6x^2 + 12x}{3x}$$
$$-3x^3 + 12x^2 - 15x + \frac{9x^3 + 6x^2 + 12x}{3x}$$

 **DEFINITION**

> The **distributive property** says that the product of a number and a sum (or difference) can be found by multiplying the number by each term of the sum (or difference). In symbols, this is $a(b + c) = ab + ac$.

Now turn your attention to $\frac{9x^3 + 6x^2 + 12x}{3x}$. Remember, the fraction bar acts as both a division sign and a grouping symbol. You must divide the entire numerator by $3x$, but you can take it one term at a time—$\frac{9x^3}{3x} = 3x^2$, $\frac{6x^2}{3x} = 2x$, and $\frac{12x}{3x} = 4$—so you have the following.

$$-3x^3 + 12x^2 - 15x + \frac{9x^3 + 6x^2 + 12x}{3x}$$
$$-3x^3 + 12x^2 - 15x + \frac{9x^3}{3x} + \frac{6x^2}{3x} + \frac{12x}{3x}$$
$$-3x^3 + 12x^2 - 15x + 3x^2 + 2x + 4$$

Finally, combine like terms.

$$-3x^3 + 12x^2 - 15x + 3x^2 + 2x + 4$$

$$-3x^3 + \left(12x^2 + 3x^2\right) + \left(-15x + 2x\right) + 4$$

$$-3x^3 + 15x^2 - 13x + 4$$

CHECKPOINT

Simplify each expression.

6. $6 - 8 + 4(11 - 3) \div 16$

7. $9x - 15x \div 3 - 7x + 13x$

8. $-4(7y - 8) + 11y - (5 - 6y)$

9. $\frac{42x+18}{6} - \frac{18(12-28x)}{2^3}$

10. $\frac{9\left(40x-75x^2\right)}{15x}$

Solving Equations and Inequalities

The primary focus of algebra is solving equations and inequalities, but doing that efficiently requires you to simplify first. Much of the work of solving actually goes on before the true solving starts, simplifying each side of the equation or inequality. Solving actually begins when both sides have been simplified down to no more than two terms.

Equations and inequalities that involve absolute value or quadratics require variations on the following techniques, but those—and many others—will be covered in later chapters. For any type of equation or inequality, remember that simplifying before you start to solve will facilitate the process.

Linear Equations

For *linear equations in one variable,* there are only four possible steps in solving: add the same quantity to both sides, subtract the same quantity from both sides, or multiply or divide both sides by the same nonzero quantity.

Let's look at an example. To solve $2(x - 1) = -3(x + 2)$, you first need to use the distributive property to eliminate the parentheses on each side.

$$2(x - 1) = -3(x + 2)$$

$$2x - 2 = -3x - 6$$

Once the parentheses are eliminated and there are no more than two terms on each side, the real work of solving can begin. You want only one variable term, so eliminate the other one by adding $3x$ to both sides.

$$2x - 2 + 3x = -3x - 6 + 3x$$

$$5x - 2 = -6$$

Add 2 to both sides, and then divide both sides by 5.

$$5x - 2 + 2 = -6 + 2$$

$$5x = -4$$

$$\frac{5x}{5} = \frac{-4}{5}$$

$$x = -\frac{4}{5}$$

Linear Inequalities

For *linear inequalities,* the same steps apply as for linear equations, but there's an additional concern. If you multiply or divide both sides of an inequality by a negative number, the direction of the inequality reverses.

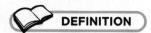 **DEFINITION**

> A **linear equation in one variable** is a mathematical sentence that says two expressions made up of constants and first-degree variable terms are equal. A **linear inequality** says that the relationship between two expressions made up of constants and first-degree variable terms is unequal, with one expression larger than the other. In either case, solving means determining the value(s) of the variable for which that sentence is true.

To solve the inequality $x + 2 + 6 > 2x + 5(x + 1)$, start in the same way as you did with the linear equation example, by simplifying the expression of each side. First, combine like terms on the left.

$$x + 2 + 6 > 2x + 5(x + 1)$$

$$x + 8 > 2x + 5(x + 1)$$

Next, clear parentheses on the right side and combine like terms.

$$x + 8 > 2x + 5x + 5$$

$$x + 8 > 7x + 5$$

With no more than two terms on a side, it's time to solve. Eliminate one of the variable terms by subtracting $7x$ from both sides.

$$x + 8 - 7x > 7x + 5 - 7x$$

$$-6x + 8 > 5$$

Subtract 8 from both sides.

$$-6x + 8 - 8 > 5 - 8$$

$$-6x > -3$$

Divide both sides by -6, however remember that this is an inequality, so dividing by a negative number will reverse the direction of the inequality.

$$\frac{-6x}{-6} > \frac{-3}{-6} \text{ becomes } x < \frac{1}{2}$$

CHECKPOINT

Solve each equation or inequality, simplifying before solving.

11. $4x - 2(3x - 8) = 12 - 7(3 - 4x)$

12. $2 - (7x + 5) = 13 - 3x$

13. $3(4x + 2) = 8 - (2x + 12) + x - 3$

14. $2 - 5(x - 3) < 1 + x - 8$

15. $x - (8 - 2x) \geq 6 + 5x - 18$

Graphing Lines

Another topic from algebra I that you'll call upon repeatedly is graphing linear equations. The graph of any equation is made up of all the points whose coordinates solve the equation, and therefore is a picture of the solution set. The first and fundamental method of constructing any graph is to build a table of values by choosing values for x and evaluating the equation to find the corresponding value of y until enough of a pattern emerges to make you confident of the shape of the graph.

The equations of lines are easy to recognize. With the exception of vertical lines, a special case, the equation of a line expresses the value of y as a multiple of x plus a constant. The equation, if arranged nicely, follows the pattern $y = mx + b$. The coefficient of x, traditionally designated as m, represents the slope of the line, and the constant, b, is the y-intercept, the point where the line crosses the y-axis.

Slope

The *slope* of the line is a rate of change. It compares the amount the line rises (or falls) to its horizontal movement: slope=$\frac{\text{rise}}{\text{run}}$. For example, if a line has the equation $y=\frac{3}{2}x-1$, it has a slope equal to $\frac{3}{2}$, which means that it rises 3 units each time it moves 2 units right.

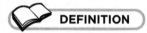

DEFINITION

The **slope** of a line is the ratio of the number of spaces up or down to the number of spaces horizontally you must count to get from one point on the line to another.

If any two points on a line are known, the slope can be calculated. The rise is the difference of the y-values, and the run is the difference of the x-coordinates. If the two points are designated as (x_1, y_1) and (x_2, y_2), the slope is $m=\frac{y_2-y_1}{x_2-x_1}$. So if a line passes through the points (–4, 3) and (6, 8), the slope of the line is $m=\frac{8-3}{6-(-4)}=\frac{5}{10}=\frac{1}{2}$. A positive slope indicates that the line rises from left to right, a negative slope shows a drop from left to right, and a 0 slope—no rise—means the line is horizontal.

Finding the Points for a Line

The equation of a line is not always arranged in a way that reveals its slope and y-intercept, but it is usually possible to isolate y with a little algebra. Even if the equation is arranged differently, if there's an x and a y without exponents, you're looking at a linear equation.

When you recognize an equation as a linear equation, there are two quick methods of drawing its graph. If the equation is arranged in or can easily be put in slope-intercept form ($y = mx + b$), the y-intercept, b, gives you a starting point on the y-axis, and the slope, m, lets you count to another point on the line. For instance, to graph the equation $y=-\frac{2}{3}x+7$, start at 7 on the y-axis, placing a dot there, and then move down 2 and 3 to the right and place another dot. Repeat the slope a few times. Once you've identified a few points, you can connect them into a line, as you can see in the following graph.

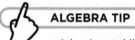

ALGEBRA TIP

A horizontal line never rises ($m = 0$), so it has an equation of just $y = b$. A vertical line also has a different type of equation because it rises but does not run, so its slope is undefined. Any number representing a rise over a 0 in the denominator produces a fraction that is undefined. A line with an undefined slope can't fit the $y = mx + b$ form, so a vertical line has an equation of the form $x =$ a constant.

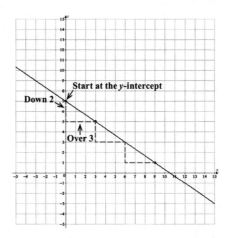

If the equation is in standard form, $ax + by = c$, you can easily find both the x-intercept and the y-intercept. If you're asked to graph $5x - 2y = 10$, for example, you can find the x-intercept by replacing y with 0 and solving for x: $5x - 2(0) = 10$ becomes $5x = 10$ or $x = 2$. To find the y-intercept, replace x with 0: $5(0) - 2y = 10$, which gives you $y = -5$. Plot the x-intercept and the y-intercept, connect them, and extend the line, as shown in the graph.

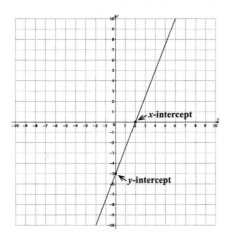

If an equation is not arranged in either of these forms, you can use algebraic operations to rearrange the equation until it conforms to either slope-intercept or standard form.

CHECKPOINT

Graph each line.

16. $y = \frac{3}{5}x - 4$

17. $y = -3x + 8$

18. $6x + 15y = 30$

19. $9x - y = -9$

20. $5y - 20 = 4x$

Writing a Linear Equation

While you're often given an equation and asked to graph it, sometimes you need to go in the other direction. If you have a graph of a line or some information about a line and you need to find the equation of the line, you need two pieces of information: the slope and a point on the line, or two distinct points on the line.

ALGEBRA TIP

If you know the slope and the y-intercept of a line, you can use the slope-intercept form of the linear equation, $y = mx + b$, rather than point-slope form. Just put the slope and y-intercept in the appropriate spots.

If you know the slope and some point on the line, you can start with point-slope form: $y - y_1 = m(x - x_1)$. For example, to write the equation of a line with a slope of $\frac{3}{5}$ that passes through the point $(-5, -8)$, start with $y - y_1 = m(x - x_1)$. The x and y without subscripts will stay as they are. Replace x_1 with the x-coordinate of your point and y_1 with the y-coordinate: $y - (-8) = m(x - (-5))$. The slope you're given goes in place of m: $y - (-8) = \frac{3}{5}(x - (-5))$. Technically, this is the equation of the line, but it's not the cleanest form, so take a moment to simplify.

$$y - (-8) = \frac{3}{5}(x - (-5))$$
$$y + 8 = \frac{3}{5}(x + 5)$$
$$y + 8 = \frac{3}{5}x + 3$$
$$y = \frac{3}{5}x - 5$$

If you know two points on the line, you have an extra step. You need to use the slope formula, $m = \frac{y_2 - y_1}{x_2 - x_1}$, to find the slope of the line. You then can use that slope and either one of the points— your choice—to plug into point-slope form. For instance, the line that passes through $(-4, 5)$ and $(4, -3)$ has a slope of $m = \frac{-3-5}{4-(-4)} = \frac{-8}{8} = -1$. In point-slope form $y - y_1 = m(x - x_1)$, replace m with

−1 and choose one of the points to replace x_1 and y_1. If you choose (−4, 5), you get $y - 5 = -1(x - (-4))$. If you choose (4, −3), it looks like this: $y - (-3) = -1(x - 4)$. Both simplify to $y = -x + 1$.

$$y - 5 = -1\left(x - (-4)\right) \qquad y - (-3) = -1\left(x - 4\right)$$
$$y - 5 = -1\left(x + 4\right) \qquad y + 3 = -1\left(x - 4\right)$$
$$y - 5 = -x - 4 \qquad y + 3 = -x + 4$$
$$y = -x + 1 \qquad y = -x + 1$$

CHECKPOINT

Find the equation of the line described.

21. slope = $\frac{1}{4}$ and y-intercept (0, −5)

22. slope = −2 through the point (5, −1)

23. slope = $\frac{2}{3}$ through the point (−6, −5)

24. through the points (−3, 1) and (5, 5)

25. through the points (−2, −7) and (2, 1)

FOIL and Factoring

Solving linear equations and inequalities are crucial algebraic skills, but another important piece of the algebraic foundation is work with polynomials. A polynomial is a sum of terms, each of which is a product of a coefficient and an integer power of a variable.

Addition, subtraction, and multiplication of polynomials are applications of the distributive property and combining like terms, but probably the most common form of polynomial multiplication is the multiplication of two binomials (polynomials with two terms).

For that multiplication, you likely learned the mnemonic FOIL: First, Outer, Inner, Last. Each of those four words denotes a step in the multiplication process. For example, to multiply $(x + 4)(x + 5)$, you multiply as follows:

FIRST: $x \cdot x = x^2$

OUTER: $x \cdot 5 = 5x$

INNER: $4 \cdot x = 4x$

LAST: $4 \cdot 5 = 20$

Combine like terms, which often are in the inner and outer positions, to get $x^2 + 9x + 20$.

When the binomials contain both positive and negative terms, the multiplication is done the same way; however, you need to be attentive to signs. For example, to multiply $(x - 7)(x + 3)$, multiply as follows:

FIRST: $x \cdot x = x^2$

OUTER: $x \cdot 3 = 3x$

INNER: $-7 \cdot x = -7x$

LAST: $-7 \cdot 3 = -21$

Combine like terms, with particular attention to signs, to get $x^2 - 4x - 21$.

If the variables have coefficients other than 1, the multiplication gets a little more involved, but the process is the same.

$$(3x - 8)(5x + 7) = 3x \cdot 5x + 3x \cdot 7 - 8 \cdot 5x - 8 \cdot 7$$
$$= 15x^2 + 21x - 40x - 56$$
$$= 15x^2 - 19x - 56$$

CHECKPOINT

Multiply each pair of binomials.

26. $(x + 7)(x + 4)$

27. $(x - 3)(x - 5)$

28. $(x - 6)(x + 9)$

29. $(2x + 5)(3x - 7)$

30. $(4x - 7)(3x + 2)$

Factoring a Quadratic Trinomial

Understanding how the FOIL rule for multiplication works not only simplifies multiplication but also helps you to factor. Factoring is the process of rewriting a polynomial as the product of two polynomials.

Like multiplication, factoring comes in many forms, but the most common type is expressing a *quadratic trinomial,* like $x^2 - 4x - 21$ or $15x^2 - 19x - 56$ as a product of two binomials. This type of factoring amounts to applying the FOIL rule in reverse, so understanding the FOIL rule gives you an advantage.

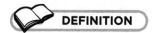 **DEFINITION**

> A **quadratic trinomial** is an algebraic expression with three terms. It contains a term with the variable squared, a term with the variable to the first power, and a constant term.

To factor a trinomial in which the coefficient of the squared term is 1, like $x^2 - 7x + 12$, remember that it would have been produced by multiplying two binomials, each of which would start with a simple x: $x^2 - 7x + 12 = (x \ \ ?)(x \ \ ?)$. The numbers that replace the question marks, the lasts, will multiply to 12, so start with a list of possible factor pairs for 12: 1 and 12, 2 and 6, or 3 and 4. The inner and the outer products must combine to make the $-7x$. The best choice is $x^2 - 7x + 12 = (x - 3)(x - 4)$, and you can check by multiplying.

While factoring is an important tool, unfortunately not all quadratic trinomials are factorable. Such expressions are said to be prime polynomials.

Factoring When the Coefficient of x^2 Isn't 1

Factoring when the coefficient of x^2 is some number other than 1 takes a bit more trial and error. Start with two different lists of possible factors: one for possible factors of the constant, and another for possible factors of the squared term. You may be fortunate enough to find that one of those coefficients is prime, and therefore has only one possible set of factors. For instance, to factor $3x^2 + 22x - 45$, start with the factors of 45—which are 1 and 45, 3 and 15, or 5 and 9—and the factors of the 3 from $3x^2$. The number 3 is prime, so its only factors are 1 and 3.

Now start the search for factors of $3x^2 + 22x - 45$ by setting up $(3x \ \ ?)(x \ \ ?)$. Choose one possible factor pair for 45 and put the numbers in place of the question marks. Let's start with 1 and 45: $(3x \ \ 1)(x \ \ 45)$. Check the inner and the outer to see if they can combine to make $22x$. Unfortunately, $1x$ and $135x$ will neither add nor subtract to $22x$, but before you abandon a factor pair, switch their positions. Change to $(3x \ \ 45)(x \ \ 1)$ and check again. $45x$ and $3x$ are not as far off, but they still don't work. Now you can cross 45 and 1 off your list and try something else, like 3 and 15 or 5 and 9, but remember to switch positions before you rule a pair out.

$(3x \ \ 3)(x \ \ 15)$ gives you $45x$ and $3x$

$(3x \ \ 15)(x \ \ 3)$ gives you $9x$ and $15x$

$(3x \ \ 5)(x \ \ 9)$ gives you $27x$ and $5x$

$(3x \ \ 9)(x \ \ 5)$ gives you $15x$ and $9x$

The only attempt that could give you a middle term of $22x$ is $(3x \ \ 5)(x \ \ 9)$, but you still need to place the signs. To produce -45, you'll need a negative number and a positive number, but do you want $(3x + 5)(x - 9)$ or $(3x - 5)(x + 9)$? Look again at the inner, $5x$, and the outer, $27x$.

To produce $+22x$, you'll need $+27x$ and $-5x$, and only the second possibility, $(3x - 5)(x + 9)$, will give you that.

CHECKPOINT

Factor completely.

31. $x^2 - 6x + 5$

32. $x^2 + 11x + 30$

33. $x^2 + x - 20$

34. $6x^2 + 13x + 7$

35. $15x^2 - 4x - 4$

The Least You Need to Know

- The real numbers are made up of the rationals, which can be written as fractions, and irrationals, which appear most often as square roots or nonrepeating, nonterminating decimals.

- Follow the order of operations, use the distributive law, and combine like terms to simplify expressions.

- Solve equations and inequalities by performing inverse operations until the variable is isolated. For inequalities, multiplying or dividing by a negative number reverses the inequality sign.

- Graph a linear equation by plotting the y-intercept and counting out the slope, or by plotting both the x-intercept and the y-intercept, and connecting.

- To write the equation of a line, find the slope and at least one point on the line, place them in $y - y_1 = m(x - x_1)$, and simplify.

- Multiply binomials by performing the four multiplications denoted by FOIL and combining like terms. Factor quadratic trinomials by reversing the FOIL rule and finding factors for the square term and for the constant term, which will produce inner and outer products that combine to the correct middle term.

Functions

As you learned in the previous chapter, a first course in algebra is aimed primarily at equipping you with the essential skills to simplify expressions, solve equations and inequalities, and draw and interpret graphs. With those tools, you can move on to a wider view of the world of algebra, and that includes an examination of various families of functions and some relations that are not functions. To do that, you first need a clear understanding of functions and how they work, which is what you'll find in this chapter.

Relations and Functions

In common speech, when we talk about relations and relationships, we're usually talking about two people or things that are somehow connected or that influence one another in some way. In math, a *relation* is defined as a pairing of numbers from one set (input) called the *domain*, with numbers from another set (output) called the *range*. For example, if you pair the numbers in the set {1, 2, 3} with the numbers in the set {6, 7, 8}, you create a relation; (1, 6), (2, 7), (3, 8) is a relation.

In This Chapter

- Understanding functions and function notation
- Creating families of functions by transformations
- Creating new functions by combining others
- Undoing the work of a function with an inverse function

If you notice, the pairing in the example seems to follow a rule. Each number from the domain is paired with a number from the range that is 5 units larger: 6 is 5 more than 1, 7 is 5 more than 2, and 8 is 5 more than 3. While sometimes relations have a rule like this to follow, other times they may not. For instance, the pairing (1, 7), (2, 6), (3, 8) is also a relation, but it has no particular pattern. A relation may match each number from the domain with a different number from the range, repeat numbers from the range, or match the same number from the domain with more than one number from the range. Every number in the domain gets a partner, and the range is the set of all those partners. For instance, if a relation matched people's ages with their annual salary, people of different ages might earn the same salary, so different elements of the domain (ages) would be paired with the same element of the range (salary). That can happen in a function. If two people the same age had different salaries, that would mean the same input had two different outputs; that's fine for a relation, but not for a function.

A *function* is a relation—a pairing—in which each number in the domain has only one partner from the range. While numbers from the range can be repeated, numbers from the domain cannot. The pairing (1, 6), (2, 6), (3, 8) is an example of a function, because each of the numbers 1, 2, and 3 have just one partner; however, the pairing (1, 6), (1, 7), (2, 8), (3, 6) can be classified as a relation but is not a function. Think of a function as a machine. You put a number from the domain into the machine, and it whirs and clangs a bit, and then turns out a number from the range.

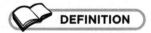 **DEFINITION**

> A **relation** is the pairing of numbers from one set with numbers from another set. The set of all values that can be substituted in a function for x—that is, all possible inputs—is called the **domain.** The set of outputs, all of which are values of y, is the **range.** A **function** is a relation in which each number in the domain has only one partner from the range.

While functions, like relations, don't have to follow a rule, the ones that do have rules are usually the most interesting and therefore the ones you'll look at most often. To describe a function properly, you need both the rule (if it has one) and the domain. For instance, knowing that a function squares the inputs it is given is good, but you also need to know what inputs it will accept: Positive numbers? Integers? All real numbers? If no domain is specified, you can assume it is all real numbers.

Function Notation

Functions are usually designated by a letter name, like f or g, followed by a set of parentheses containing the variable that represents the elements of the domain. So a function that matches each number in the domain with a number 3 units larger might be represented as $f(x) = x + 3$, while a function that matches each number in the domain with its square could be written as

$g(x) = x^2$. This function notation allows you to refer to functions by their "names"—f, g, or some other letter—rather than go through the whole description of the function every time you want to refer to it. It announces that the relationship you're describing is, in fact, a function, and it gives you a quick way to say "Find the value from the range that should be paired with this value from the domain."

For example, if your function is $g(x) = x^2$ and you want to ask for the value that gets paired with 3, you can simply ask for $g(3)$. Because $g(x) = x^2$ and you're replacing the x with a 3, $g(3) = 3^2 = 9$. As you can see, the number 3 is paired with 9. Or if you're using the function $f(x) = x + 3$, $f(-5)$ says to replace the x with -5, so $f(-5) = -5 + 3 = -2$. While functions may be defined by simple rules or complicated ones, the function notation tells you how to evaluate the function no matter what the rule may be.

 ALGEBRA TIP

If you have the graph of a relation, you can quickly tell if the relation is a function by doing a vertical line test. Hold your pencil so it creates a vertical line over the graph, and move it across the graph from left to right. If it ever crosses the graph twice, the relation is not a function.

Evaluating functions can be useful in cases such as calculating a shipping cost. If you know that the shipping company you want to use to deliver products to your online customers charges $8.10 for the first pound plus 47¢ for each additional pound, you can write a rule for the shipping cost as a function of the package weight: $C(p) = 8.10 + 0.47p$. The domain of the function, p, would be positive whole numbers because the shipping company rounds to the nearest pound. That function would allow you to calculate the cost of any shipment you were planning.

Building New Functions

Algebra is interested in the practical and the theoretical use of functions, and one important topic is building new functions from known functions. Sometimes this building process is the result of the situation in which you're using the functions. For example, if one function calculates the surface area of the walls in a room, another function tells you how much paint is needed per square foot of wall, and a third determines the cost of the paint, they all work together to estimate what it will cost to paint the room. But in a more theoretical sense, if you can understand how a function is built from simpler ones, you can more easily understand what its domain and range may be, what its graph will look like, and how it will behave.

One way to create new functions from old ones is to take a function like $f(x) = x^2 - 5x + 9$ and evaluate it, not for a constant, but for a variable expression. If you write $f(2t - 1)$, you replace every x in the function rule with $2t - 1$.

$$f(2t-1)=(2t-1)^2-5(2t-1)+9$$
$$=4t^2-4t+1-10t+5+9$$
$$=4t^2-14t+15$$

You started with a function whose inputs were represented by the variable x and produced a new function whose inputs are denoted by the variable t. Both of these functions have domains of all real numbers, but sometimes the function produced will have a different domain than the one you started with. You often see that when you build new functions by adding, subtracting, multiplying, or dividing functions.

CHECKPOINT

Determine if the relation is a function.

1.

x	-3	-1	0	1	3
y	4	-4	-5	-4	4

2. $x^2 + 4y^2 = 16$

Evaluate each function.

3. For $f(x) = 2x^3 - 8x^2 + x - 5$, find $f(-1)$.

4. For $g(x) = \sqrt{5x^2 - 4}$, find $g(2)$.

5. For $f(x) = \frac{2x-8}{x+1}$, find $f(0)$.

Function Arithmetic

Function arithmetic creates new functions by the operations of arithmetic: addition, subtraction, multiplication, and division.

Say you begin with the functions $f(x) = x^2 - 5$ and $g(x) = 9 - 7x$; you can form the function $f + g(x)$ by adding the rules of $f(x)$ and $g(x)$.

$$f + g(x) = f(x) + g(x) = (x^2 - 5) + (9 - 7x)$$

As usual, you'll want to simplify that as much as possible.

$$f + g(x) = x^2 - 7x + 4$$

In the same way, you can form a new function by subtraction $(f - g(x) = (x^2 - 5) - (9 - 7x) = x^2 + 7x - 14)$, multiplication $(f \cdot g(x) = (x^2 - 5) - (9 - 7x) = -7x^3 + 9x^2 + 35x - 45)$, or division $(\frac{f}{g}(x) = \frac{x^2-5}{9-7x})$.

ALGEBRA TIP

To find a sum, difference, product, or quotient of two functions for a specific value of x, you can evaluate each function for that value and do the arithmetic on the outputs. So if $f(x) = x^2 - 5$ and $g(x) = 9 - 7x$, $f + g(-2) = f(-2) + g(-2) = -1 + 23 = 22$.

Because you're working with functions, you'll also want to stop a moment and think about domains. Whenever you divide, the function in the denominator cannot equal 0, so for $\frac{f}{g}(x) = \frac{x^2-5}{9-7x}$, $9 - 7x \neq 0$; this means $x \neq \frac{9}{7}$. For addition, subtraction, and multiplication, in this case, both f and g are defined for all real numbers, so the new function, $f + g$, will have a domain of all reals as well.

Sometimes, however, the domain can't be all real numbers. For instance, the function $r(t) = \sqrt{t + 3}$ has a limited domain, because in the real numbers, you cannot find the square root of a negative number. The expression under the square root sign, the radicand, must be greater than or equal to 0, and $t + 3 \geq 0$ means t must be greater than or equal to -3. The function $p(t) = \sqrt{t - 5}$ has a domain of $t \geq 5$ for the same reason. When you start to do arithmetic with these two functions, you have to find a domain on which both of them exist. That's where the two domains overlap—in this case, $t \geq 5$, as you can see in the following illustration.

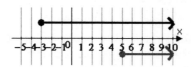

The function would be formed by arithmetic, just as before, although there won't be much simplifying possible. $r + p(t) = \sqrt{t + 3} + \sqrt{t - 5}$, $r - p(t) = \sqrt{t + 3} - \sqrt{t - 5}$, and $r \cdot p(t) = \sqrt{t + 3} \cdot \sqrt{t - 5} = \sqrt{t^2 - 2t - 15}$ will all have a domain of $t \geq 5$. The function created by dividing, $\frac{r}{p}(t) = \frac{\sqrt{t+3}}{\sqrt{t-5}} = \frac{\sqrt{t+3}}{\sqrt{t-5}} \cdot \frac{\sqrt{t-5}}{\sqrt{t-5}} = \frac{\sqrt{t^2-2t-15}}{\sqrt{t-5}}$, will have a domain of $t > 5$ but not equal to 5, because the denominator must not equal 0.

CHECKPOINT

Given functions $f(x) = 3x - 7$ and $g(x) = 9x^2 - 49$, find each of these new functions and give the domain of the new function.

6. $f + g(x)$

7. $g - f(x)$

8. $f \cdot g(x)$

9. $\frac{f}{g}(x)$

10. $\frac{g}{f}(x)$

Composition

Another way that functions can be combined to make new functions is by composition. Composition can be imagined as one function passing information to another. The symbol $f \circ g(x)$, which you read as "f of g of x," means that an input, an x, is given to function g. Function g produces an output, which is given to function f. Function f then does its work and produces the final output. $f \circ g(x)$ can also be written as $f(g(x))$, and that notation gives you a better idea of how to create the new function.

If you simply need to evaluate the *composite function* $f \circ g(x)$, you can just work your way through the two functions. So if $f(x) = \sqrt{2x - 1}$ and $g(x) = x^2 - 4$ and you want to find $f \circ g(-3)$, you can find $g(-3) = (-3)^2 - 4 = 5$ and then evaluate $f(5) = \sqrt{2(5) - 1} = \sqrt{9} = 3$.

DEFINITION

A **composite function** is a new function formed by two functions working in sequence. For instance, $f \circ g(x)$ is f following g and $g \circ f(x)$ is g following f.

That's fine if you only need a few values, but to find the rule and the domain for the new function $f \circ g(x)$, you'll need to work a bit of algebra with the rules of the two functions. Working with the functions $f(x) = \sqrt{2x - 1}$ and $g(x) = x^2 - 4$, the new function formed by composing f and g is $f \circ g(x) = f(g(x)) = f(x^2 - 4) = \sqrt{2(x^2 - 4) - 1} = \sqrt{2x^2 - 9}$.

To find the domain of the new function, you'll need to first look at the domains of the two individual functions. The function $f(x) = \sqrt{2x - 1}$ must have $2x - 1 \geq 0$, so x must be greater than or equal to $\frac{1}{2}$. The domain of $g(x) = x^2 - 4$ is all real numbers. When you form the composition $f \circ g(x) = f(g(x))$, you begin with the domain of the first or inner function, $g(x)$, so you start, in this case, with all real numbers. You are going to pass the outputs of g to the function f, so you need to make sure everything g produces is greater than or equal to $\frac{1}{2}$. Therefore, you need to be certain $x^2 - 4 \geq \frac{1}{2}$. Isolating x^2 gives you $x^2 \geq \frac{9}{2}$, and some careful thought will tell you that x must be greater than or equal to $\frac{3}{\sqrt{2}}$ or less than or equal to $-\frac{3}{\sqrt{2}}$. If you give $g(x) = x^2 - 4$ values of x that are greater than or equal to $\frac{3}{\sqrt{2}}$ or less than or equal to $-\frac{3}{\sqrt{2}}$, it will produce values that are greater than or equal to $\frac{1}{2}$ and will be able to pass those outputs to f, so the composite function will exist.

In many cases like the example I've given, you can look at the simplified version of the composite function and identify the necessary restrictions, but not in every case. This makes it wise to think about the domain of a composite function with two rules:

1. Begin with the domain of the inner function.

2. Add any other restrictions necessary so all the outputs of the inner function are in the domain of the outer function.

Going by these rules, notice that $g \circ f(x) = g(f(x))$ is different from $f \circ g(x) = f(g(x))$, both in the function rule and the domain. If you reverse the composition order of the previous example, $g \circ f(x) = g(f(x)) = g(\sqrt{2x-1}) = \sqrt{2x-1}^2 - 4 = 3(2x-1) - 4 = 6x - 7$. So you begin with the domain of f, $x \geq \frac{1}{2}$, and then consider whether everything in the range of f is part of the domain of g. Because g has a domain of all reals, you don't need any other restrictions, so the domain of $g \circ f(x) = g(f(x))$ is $x \geq \frac{1}{2}$. As you can see, changing the order of the composition can give you a very different function.

ALGEBRA TIP

To evaluate a composite function for a specific value of x, you do not necessarily need to find the new rule. You can evaluate the first function for the specified value and pass that output to the second function. If $f(x) = \sqrt{2x-1}$ and $g(x) = x^2 - 4$, to evaluate $f \circ g(3) = f(g(3))$, first find $g(3) = (3)^2 - 4 = 5$, and then pass the 5 to the function f:
$f(5) = \sqrt{2(5) - 1} = \sqrt{9} = 3$. So $f \circ g(3) = f(5) = 3$.

CHECKPOINT

Given $f(x) = \sqrt{x-2}$, $g(x) = x^2 + 2$, and $h(x) = 3x - 1$, find each composition and give its domain.

11. $f \circ g(x)$

12. $g \circ f(x)$

13. $f \circ h(x)$

14. $g \circ h(x)$

15. $h(f(g(x)))$

Inverses

Much of the work in algebra is concerned with undoing some operation that had been done. Solving an equation involves stripping away all the operations that have been performed on a variable to get back to the value of the unknown. That's accomplished by performing the opposite, or inverse, operations.

The algebra of functions also contains a concept of an inverse. Two functions are inverses if the composition of the two, in either order, is equal to x. Two functions, $f(x)$ and $g(x)$, are inverses if $f(g(x)) = g(f(x)) = x$. The job of an inverse function is to undo the work of another function. If $f(x) = x + 2$, its job is to add 2 to whatever it's given. Its inverse is a function that subtracts 2 from every input, or $g(x) = x - 2$. When you compose the two functions, $f \circ g(x) = x - 2 + 2 = x$ and $g \circ f(x) = f(x+2) = x + 2 - 2 = x$.

Of course, the functions aren't usually that simple, but if you want to know if two functions, f and g, are inverses, you simply need to check that $f(g(x)) = g(f(x)) = x$. For example, if $f(x) = 3x + 7$ and $g(x) = \frac{1}{3}x - \frac{7}{3}$, the composition $f(g(x)) = f\left(\frac{1}{3}x - \frac{7}{3}\right) = 3\left(\frac{1}{3}x - \frac{7}{3}\right) + 7 = x - 7 + 7 = x$ and the composition $g(f(x)) = g(3x + 7) = \frac{1}{3}(3x + 7) - \frac{7}{3} = x + \frac{7}{3} - \frac{7}{3} = x$, so the functions are inverses. You can denote the inverse of $f(x)$ as $f^{-1}(x)$ or the inverse of $g(x)$ as $g^{-1}(x)$. In the case of the example, if $f(x) = 3x + 7$, $f^{-1}(x) = \frac{1}{3}x - \frac{7}{3}$; if $g(x) = \frac{1}{3}x - \frac{7}{3}$, $g^{-1}(x) = 3x + 7$.

ALGEBRA TRAP

The word *inverse* can mean "opposite" in many different circumstances. The symbol for inverse function, with the superscript –1, may remind you of the –1 exponent that signifies a reciprocal. Multiplying a number by its reciprocal produces a product of 1, while composing a function with its inverse function results in x, the identity function. The ideas are similar, but one talks about multiplication and the other about composition of functions. Don't confuse the inverse of a function with its reciprocal.

Unfortunately, you aren't always going to have both a function and its inverse. Sometimes, you're given a function and asked to find its inverse function. The first thing you need to consider is the function you're given. Not every function has an inverse function. Functions match each input with only one output, but in many cases, outputs are matched to more than one input. The squaring function, $f(x) = x^2$, for example, matches the input 2 with the output 4, but also matches the input –2 with the same output. That's fine for the function, but it means the inverse will not be a function. The inverse would need to pair an input of 4 with both 2 and –2, but a function can only give an input one partner.

So before you go looking for an inverse function, you have to make sure the inverse will in fact be a function by making sure that the original function is one-to-one (often written as 1-1). That means each input has only one output and each output traces back to only one input. If necessary, you can restrict the domain of the function to assure that the function is 1-1. For example, if your function is $f(x) = x^2$, you might restrict the domain to only consider the function on the set of all $x \geq 0$ instead of on the entire set of real numbers. By eliminating the negative numbers, you can be certain that each output traces back to only one input.

ALGEBRA TIP

Just as you can tell whether a graph represents a function by using a vertical line test, you can check if a function is 1-1 by doing a horizontal line test. If any horizontal line crosses a graph more than once, the function is not 1-1.

To find the rule for the inverse of a function, do the following:

1. Write the function rule, replacing the function notation $f(x)$ with a y.

2. Swap the x and y. Do this carefully, especially if there is more than one appearance of x. Each x must be replaced with a y, but there should be only one y in the original, which will become an x.

3. Isolate y on one side of the equation, leaving only x-terms and constants on the other side.

To find the inverse of $f(x) = \frac{x}{x+2}$, first sketch its graph and use the horizontal line test to check that the function is 1-1. Although the graph has an unusual shape, any vertical line will intersect the graph only once. That tells you the function is 1-1, so it is possible to find an inverse function.

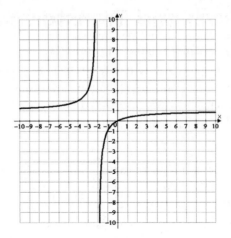

Now write the function as $y = \frac{x}{x+2}$ and swap the variables: $x = \frac{y}{y+2}$.

Solve for y by cross-multiplying and collecting all terms involving y on one side.

$$x = \frac{y}{y+2}$$
$$x(y+2) = y$$
$$xy + 2x = y$$
$$xy - y = -2x$$

If there is more than one term involving y, factor out the y, and then divide both sides by the other factor.

$$xy - y = -2x$$
$$y(x-1) = -2x$$
$$y = \frac{-2x}{x-1}$$

This equation is the rule for the inverse of $f(x) = \frac{x}{x+2}$, so you can write it as $f^{-1}(x) = \frac{-2x}{x-1}$.

CHECKPOINT

Find the inverse of each function and give its domain.

16. $f(x) = 4x - 1$

17. $g(x) = \sqrt{x^2 + 9}, x \leq 0$

18. $f(x) = 5 - x^2, x \geq 0$

19. $g(x) = \frac{1}{x}, x \neq 0$

20. $f(x) = \frac{x+1}{x-1}$

Parent Functions and Transformations

In algebra II, you explore a number of families of functions, each with its own rule and its own characteristic graph. The term *parent function* refers to the simplest member of a family, the rule and graph that can be modified easily to create all the other members of the family. The following table shows some of the common parent functions, but as you look at each of the families, you'll focus more closely on that family's parent function.

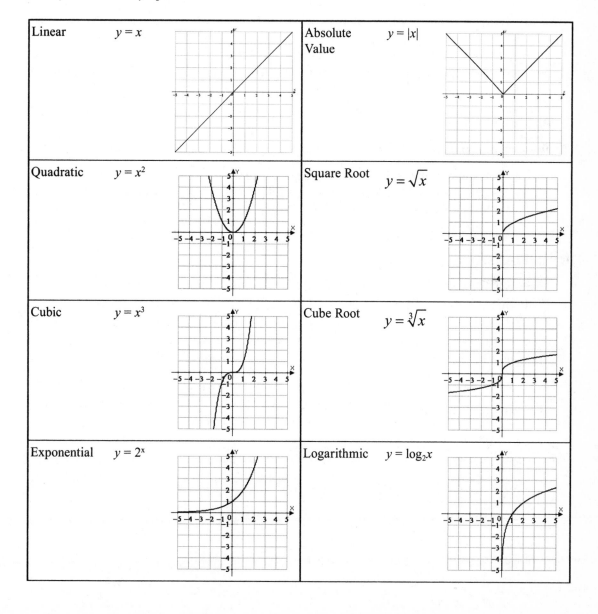

Linear	$y = x$		Absolute Value	$y = |x|$
Quadratic	$y = x^2$		Square Root	$y = \sqrt{x}$
Cubic	$y = x^3$		Cube Root	$y = \sqrt[3]{x}$
Exponential	$y = 2^x$		Logarithmic	$y = \log_2 x$

| Rational | $y = \frac{1}{x}$ | 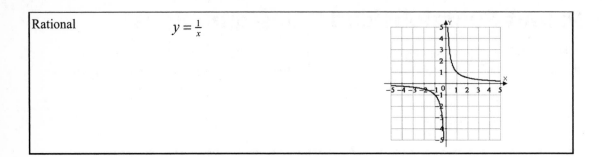 |

Many of the functions in any family are created by performing one or more *transformations* on the *parent graph*. The parent can be moved up, down, left, or right and can be flipped horizontally or vertically. The graph can also be stretched or compressed, either in the vertical or horizontal direction.

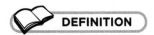 **DEFINITION**

> **Transformations** are changes made to a parent function to translate or slide it, reflect or flip it, and stretch or compress it. A **parent graph** is the simplest graph that meets the description of a family of functions.

Rigid Transformations

If you know the shape of the parent graph and can recognize the ways in which those transformations show up in the equation, you can quickly find the graph of any function in the family. Most of the changes are called *rigid transformations,* meaning that the shape of the graph doesn't change—it just moves to a new location. The rigid transformations are translation, or slide, and reflection, or flip.

Anything introduced into the rule that affects the *x* creates a change in the horizontal direction. Let's use a square root graph to illustrate the changes. I'll leave the shadow of the parent function on each of the transformations, so you can see how things have changed.

$f(x) = \sqrt{x}$ Parent function

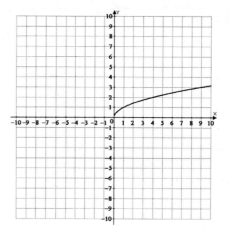

$f(x) = \sqrt{x-3}$ 3 units right

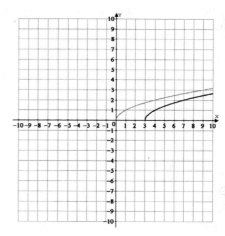

$f(x) = \sqrt{x+2}$ 2 units left

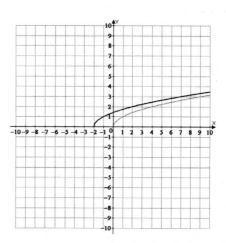

$f(x) = \sqrt{-x}$ Flipped across the y-axis

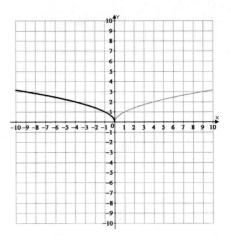

Changes made outside the square root signs—or after the work of the function—affect the y-values and cause changes in the vertical direction, as you can see in the following.

$f(x) = \sqrt{x} + 4$ 4 units up

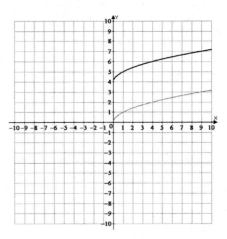

$f(x) = \sqrt{x} - 5$ 5 units down

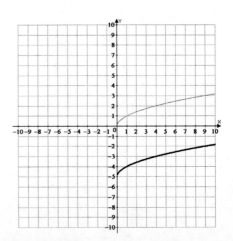

$f(x) = -\sqrt{x}$ Flipped across the x-axis

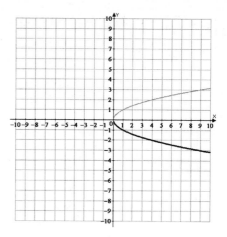

Nonrigid Transformations

The transformations that are not rigid are the stretches and compressions. These can also be horizontal or vertical, but be aware that it's often difficult to tell whether you're looking at a vertical stretch or a horizontal compression, for example. There may even be a horizontal compression and a vertical stretch that produce the same graph. Let's look at a couple of examples first.

The most common nonrigid transformation is a vertical stretch or compression. This happens when a multiplier is placed in front of the function. If the multiplier is greater than 1, the graph is stretched; if the multiplier is between 0 and 1, the graph is compressed. Each of the following examples is shown with the parent function, $f(x) = \sqrt{x}$, drawn with dotted lines, so you can compare.

$f(x) = 3\sqrt{x}$ Vertical stretch by a factor of 3

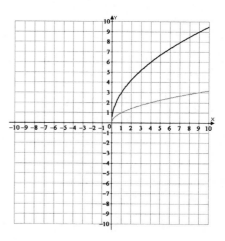

$f(x) = \frac{1}{3}\sqrt{x}$ Vertical compression by a factor of 3

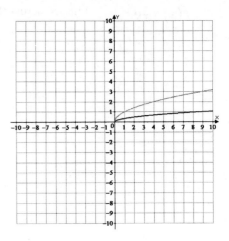

Creating Families of Functions

Families of functions are created by applying one or more transformations to a parent function. For example, the function $f(x) = -2|x + 5| - 7$ is created when the parent function for the absolute value function $y = |x|$ is shifted 5 units left, reflected over the x-axis, stretched by a factor of 2, and shifted down 7 units, as you can see in the following graph.

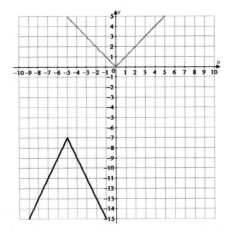

CHECKPOINT

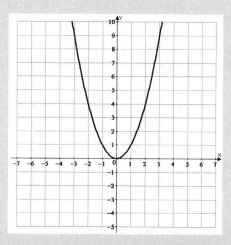

Given the function $f(x) = x^2$ whose graph is shown, sketch a graph of each of the following.

21. $g(x) = (x - 3)^2 + 2$

22. $g(x) = (x + 4)^2 - 1$

23. $g(x) = -\frac{1}{2}x^2 + 5$

24. $g(x) = -(x + 5)^2 + 3$

25. $g(x) = -2(x + 4)^2 + 7$

The Least You Need to Know

- A function is a pairing of numbers from a domain set with numbers from a range set that matches each number from the domain with only one number from the range. A function can often be defined by a rule or equation, but it should identify the domain as well.

- When the $f(x)$ notation replaces x with a value, it indicates that the function rule should be evaluated by replacing each x with that value.

- New functions can be formed by combining functions through addition, subtraction, multiplication, or division. However, domains must be restricted so that both functions exist and denominators are not 0.

- Composition creates a new function by having one function follow another— $f \circ g(x) = f(g(x))$ and the domain of the domain of g, with any additional restrictions so the range of g is within the domain of f.

- Inverse functions undo the work of one another. If f and g are inverses, $f(g(x)) = g(f(x)) = x$. To find the inverse of a function, first assure that the function is 1-1, swap the variables, and isolate y.

- Transforming a parent function by sliding, reflecting, stretching, or compressing will generate various members of the function family.

PART 2

Linear Relationships

Early in algebra (and in geometry, for that matter), you learned about lines. You learned what their equations looked like, how to graph those equations, and how to find the equation when you already have the graph. You also learned about equations and inequalities called *linear*, because they have traits in common with the equations of lines.

It's time to extend some of those ideas. In this part, you look at functions whose graphs are not simple lines, but are built from line segments and rays. You also learn about systems of linear equations and inequalities, looking at old methods and new ones made possible by matrices. Finally, just as solid geometry follows plane geometry, algebra II moves into three dimensions with systems of three equations in three variables, making those matrix methods even more welcome.

Linear Functions

In Chapter 1, you reviewed graphing linear equations and writing a linear equation based on the graph or information about the graph. Those skills from algebra I continue to be important throughout your math studies, but algebra II expands them in several ways. In this chapter, you look at graphing linear inequalities, finding the line of best fit for a collection of data points, and graphing functions that combine several different rules on different parts of their domain.

In This Chapter

- Graphing linear inequalities in two variables
- Fitting the best possible line to collected data
- Evaluating and graphing piecewise functions

Inequalities in Two Variables

Linear equations describe relationships between two variables that can be pinned down to a pair of values. When x takes this value, y must take that value to make the statement true. For example, the equation $y = 3x - 1$ tells you that when $x = 4$, y must equal 11. The ordered pair (4, 11) is a solution. And when $x = -5$, $y = -16$ according to that equation. For each proposed x-value, there is a single y-value that will make the relationship true, and so, with the exception of a vertical line, linear equations describe functions whose graphs are lines.

When the relationship is not one of equality but of inequality—of greater or less—the relationship no longer defines a function. So if $y < 3x - 1$ and you choose $x = -5$, there are many y-values that are less than -16. Each x-value has many—infinitely many—possible partners. Therefore, the graph of this kind of relation includes many more points.

Graphing an Inequality in Slope-Intercept Form

To graph a *linear inequality in two variables,* you begin with the graph of an equation. This means in order to graph $y < 3x - 1$, you first sketch the graph of $y = 3x - 1$. However, before you complete that task, you need to make a decision. Some inequalities, like $y < 3x - 1$, are strict inequalities—they want y-values to be only less than (or only greater than) some expression involving x. Other inequalities include the values of y that are equal to the expression, as well as less (or greater), which is indicated by the inequality signs $\leq$ or $\geq$. If you want to include the values that make y equal to your expression, draw the graph of the equation—the line—with a solid line. But if you are graphing a strict inequality and do not actually want the points on the line, use a dotted line. For $y < 3x - 1$, you'll use a dotted line, as you can see in the following graph.

 **DEFINITION**

A **linear inequality in two variables** is a sentence that describes a relationship between two variables in which one variable is greater than or less than an expression involving the other variable.

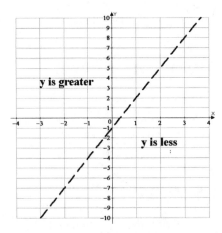

Even if you don't actually want the points on the line, you need to know where the line is, because it forms the boundary between the points you do want and the ones you don't. Every point on the line is a point that makes $y = 3x - 1$. On one side of the line are all the points for which $y < 3x - 1$, and on the other side, the points that make $y > 3x - 1$. When y is isolated, as it is in $y < 3x - 1$, you can determine which side of the line you want easily. If you want y to be less than the expression, the points that solve that inequality are below the line. If you want y to be greater, your points are above the line. So once the line has been drawn as a border, shade the side of the line that contains the points that make your inequality true. In the case of $y < 3x - 1$, you shade below the line, as shown in the following graph.

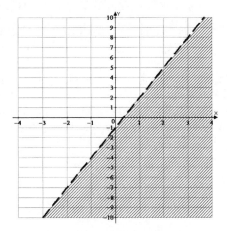

Graphing an Inequality in Standard Form

If your inequality is in standard form rather than slope-intercept form, you may choose to sketch the graph of the line by using the x- and y-intercepts. So to graph $3x - 4y \geq 12$, use a solid line to connect the x-intercept $(4, 0)$ to the y-intercept $(0, -3)$ and extend the line in both directions, as shown in the following graph. Don't be too quick to shade, because y is not isolated.

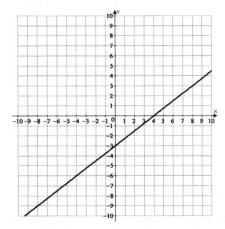

If you go through the process of isolating y, you'll have to divide both sides by -4, and that will change the direction of the inequality sign. If you don't want to go through that process, the simplest way to make sure you're shading the correct side of the line is to choose a point on one side of the line and test it in the inequality. The simplest point to choose is the origin $(0, 0)$, as long as it is not on the line. In $3x - 4y \geq 12$, if you plug in 0 for x and 0 for y, you get $3(0) - 4(0) \geq 12$, which is not true, so you do not want to shade the side of the line that contains the origin. If your point makes the inequality true, shade the side of the line that contains that point. If the point you test makes the inequality false, shade the opposite side of the line. Because the origin didn't work in $3x - 4y \geq 12$, you shade below the line, as you can see in the following graph.

 ALGEBRA TRAP

The rules of algebra don't change when you shift from sentences with one variable to sentences with two variables. When isolating y, remember that if you divide both sides by a negative number, the direction of the inequality reverses.

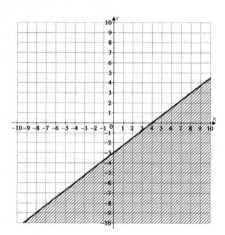

 CHECKPOINT

Graph each inequality.

1. $y \leq \frac{2}{3}x - 5$

2. $y > -4x + 9$

3. $3x - 2y < 12$

4. $2x + y \geq 8$

5. $3y < 7 - 2x$

Fitting Equations to Data

When you're presented with two points, geometry tells you there is a unique line that passes through those points, and algebra tells you how to find its slope and how to use point-slope form (see Chapter 1) to find the equation of the line. When you have a collection of points and you're looking for a single line that describes the relationship of the two variables they represent, which points do you use?

For instance, a survey of 10 people recorded their ages and annual salaries. The results are shown in the following table.

Age (years)	25	32	38	42	45	45	51	57	60	67
Salary (thousands of dollars)	30	38	47	45	55	50	48	65	110	85

If you plot these points with age as x and salary as y, shown in the following graph, the pattern is certainly not perfect but it does suggest a linear relationship. But what is that relationship? If you randomly select two of these points and find the line that connects them, is that the best description of the relationship? Choosing different points could produce very different lines.

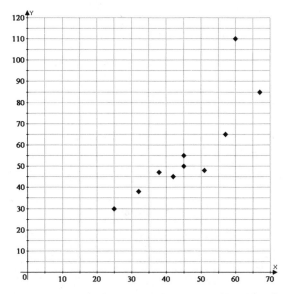

The true *line of best fit* may not actually pass exactly through any two points in the group. The idea behind the best fit line is to minimize the error—to keep the total amount by which the line misses points as small as possible—so a line that misses every point by a very small amount may be better than a line that goes through two points perfectly but misses others by a lot.

DEFINITION

If a line is fit to a set of data, the distance between the observed value of y and the value of y which the line predicts for each x-value is a residual. The **line of best fit** is the line for which the total of the squares of the residuals is a minimum.

Finding the line of best fit by hand can be a bit complicated, but the one point you can be certain the line of best fit will pass through is the point whose x-coordinate is the average of the x's and whose y-coordinate is the average of the y's. This point, which is designated $(\overline{x}, \overline{y})$, may or may not be a data point, but it is a point on the line of best fit. You could use that and one other point from the data to make your best guess at the line, but if you're willing to do a little more arithmetic, you could find the slope. A point and a slope give you enough information to write the equation of the line.

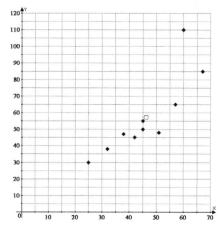

Take your data table and add two more rows—one for the product of x and y for each point, and another for the squares of the x's. Fill those in and then find the total of each row. The results are shown in the following table.

	Data Values										Total
x	25	32	38	42	45	45	51	57	60	67	462
y	30	38	47	45	55	50	48	65	110	85	573
xy	750	1,216	1,786	1,890	2,475	2,250	2,448	3,705	6,600	5,695	28,815
x^2	625	1,024	1,444	1,764	2,025	2,025	2,601	3,249	3,600	4,489	22,846

Divide the total of the x's and the total of the y's by 10—the number of data points—to find the averages. (46.2, 57.3) is a point on the line.

To find the slope, calculate as follows.

$$m = \frac{(\text{sum of the } x\text{'s})(\text{sum of the } y\text{'s}) - (\text{number of points})(\text{sum of the } xy\text{'s})}{(\text{sum of the } x\text{'s})^2 - (\text{number of points})(\text{sum of the } x^2\text{'s})}$$

$$= \frac{(462)(573) - 10(28,815)}{(462)^2 - 10(22,846)}$$

$$= \frac{264,726 - 288,150}{213,444 - 228,460}$$

$$= \frac{-23,425}{-15,016}$$

$$\approx 1.56$$

With a slope of approximately 1.56 and the point (46.2, 57.3), you can use point-slope form to write the equation of the line.

$$y - y_1 = m(x - x_1)$$
$$y - 57.3 = 1.56(x - 46.2)$$
$$y - 57.3 = 1.56x - 72.072$$
$$y \approx 1.56x - 14.77$$

You can then use this equation to graph the line of best fit, as shown in the following.

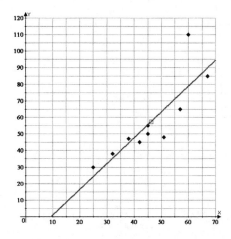

The reason to find an equation to describe the relation of x and y is usually to allow you to use the equation to predict the value of y for an x-value not already in the data set, or to calculate what x-value might produce a certain y. You might use the equation to make an estimate of the average salary of a 40-year-old, or to estimate at what age a person might earn $60,000 a year.

To estimate a salary at age 40, take the equation $y \approx 1.56x - 14.77$ and replace x with 40.

$$y \approx 1.56x - 14.77 \approx 1.56(40) - 14.77 \approx 47.63$$

Your equation predicts that a 40-year-old will earn approximately $47,630 per year. But at what age could a person expect to earn $60,000 a year? For that estimate, replace y with 60, and solve for x.

$$y \approx 1.56x - 14.77$$
$$60 \approx 1.56x - 14.77$$
$$74.77 \approx 1.56x$$
$$\frac{74.77}{1.56} \approx x$$
$$x \approx 47.93$$

With this model, you would predict an annual salary of $60,000 at approximately age 48.

CHECKPOINT

Find the line of best fit for each set of points. Use a graphing calculator or software if available.

6.

x	5	6	6	7	8	9	9	10	12
y	18	19	16	12	17	14	12	8	10

7.

x	1	3	6	7	8	12	14	15	15
y	18	32	46	52	65	101	113	111	116

8.

x	24	28	31	35	36	39	41	42	45
y	45	49	44	52	52	58	53	57	66

9.

x	1	2	2	3	4	4	5	5	6
y	-3.25	2.5	3.5	-1.75	0	-2	5.75	-1.25	-0.5

10.

x	12	15	18	20	22	25	30	35	36
y	346	313	278	247	223	185	124	71	52

Piecewise Functions

Linear equations (except for vertical lines) define functions, but linear inequalities do not. Inequalities in two variables are relations but not functions. Data sets may be functions or not, but whether the data points form a function or a relation, the line of best fit is an attempt to find one equation—one function—that describes the relationship reasonably well. Like most of your experiences with functions, it looks for a single rule.

But not every function can be defined by a single rule. Some relationships follow different rules for different values of x. A *piecewise function* (or piecewise-defined function) is a function that is determined by different rules on different parts of its domain. The sections must be disjoint, because any overlap would mean the same x might have two different partners from two rules, and the relationship wouldn't be a function.

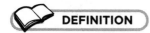 **DEFINITION**

A **piecewise function** is one which produces outputs according to different rules depending on what section of the domain the input comes from. The individual rules may be linear or nonlinear functions, or constant functions.

The following function definition is an example of a piecewise function.

$$f(x) = \begin{cases} -2x - 5 & \text{if } x \leq 1 \\ \frac{2}{3}x - 1 & \text{if } x > 1 \end{cases}$$

It tells you that if the value of x that you are passing to the function is less than or equal to 1, you should follow the rule $y = -2x - 5$ to determine the output, but when the x is greater than 1, you should find y by computing $y = \frac{2}{3}x - 1$.

You never calculate both rules for the same x-value. For instance, if $x = -3$, you first notice that -3 is less than 1, and you use the top rule: $y = -2(-3) - 5 = 6 - 5 = 1$. This function pairs -3 with 1 to make the ordered pair $(-3, 1)$. If you have an x of 6, however, you recognize that 6 is greater than 1 and so you follow the bottom rule: $y = \frac{2}{3}(6) - 1 = 4 - 1 = 3$. This gives you the ordered pair $(6, 3)$.

You need to be careful when you evaluate the function for the value of x right on the line between the two rules. One rule will apply to that value, and the other will not. To know which rule to use, check the "if" statement carefully for the "or equal to" on the sign.

Graphing a Piecewise Function

To graph a piecewise function, let's look at $f(x) = \begin{cases} 3x-2 & \text{if } x<1 \\ 5-3x & \text{if } x\geq 1 \end{cases}$. Imagine that on the left grid, you were to sketch the graph of $y = 3x - 2$, and on the right grid, sketch the graph of $y = 5 - 3x$. The results are shown in the following graphs; just ignore the dotted lines for now.

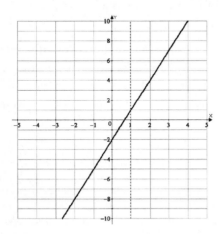
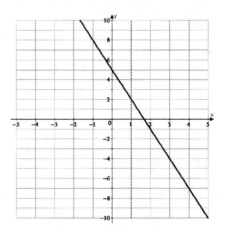

You want only part of each line because that rule is only for certain values of x. You want the portion of $y = 3x - 2$ for $x < 1$ or to the left of the dotted line $x = 1$, but you want the piece of $y = 5 - 3x$ to the right of the line, because that's the rule when $x > 1$. Imagine that you erase or cut each graph to only keep the portion you want. The graphs would look like the following.

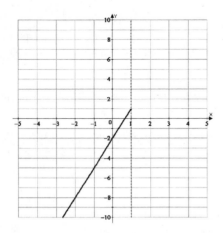
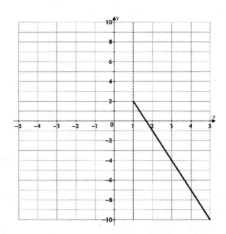

You're going to bring those two pieces together on one grid, so that you have the following graph.

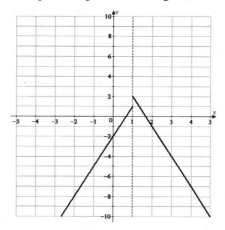

At $x = 1$, put a solid dot on the end of the section of $y = 5 - 3x$ because that is defined for $x = 1$, and an open circle on the end of the visible portion of $y = 3x - 2$, because that is only defined for values less than 1. The following is how it should look.

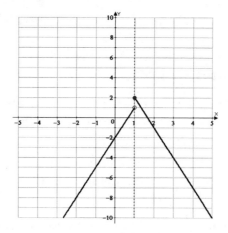

Of course, you'll want to do this on one grid and not play cut and paste, so just work lightly in pencil, so you can erase any section you don't want.

A Well-Known Piecewise Function

One of the functions you routinely study is actually a piecewise function, but because it's a commonly used function, it's been given its own name. The following piecewise function says "If x is 0 or a positive number, then y is the same number, but if x is negative, multiply it by -1, so that y will be positive."

$$f(x) = \begin{cases} x & \text{if } x \geq 0 \\ -x & \text{if } x < 0 \end{cases}$$

This is the function you know as $f(x) = |x|$—the absolute value function. You'll explore that function in the next chapter.

CHECKPOINT

Graph each piecewise function.

11. $f(x) = \begin{cases} 3x-5 & \text{if } x \geq 2 \\ 9-2x & \text{if } x < 2 \end{cases}$

14. $f(x) = \begin{cases} 2x+9 & \text{if } x < -2 \\ 5 & \text{if } -2 \leq x \leq 4 \\ x-4 & \text{if } x > 4 \end{cases}$

12. $g(x) = \begin{cases} \frac{1}{2}x+5 & \text{if } x < 0 \\ -\frac{3}{5}x-7 & \text{if } x \geq 0 \end{cases}$

13. $y = \begin{cases} x+4 & \text{if } x \geq 3 \\ x-7 & \text{if } x < 3 \end{cases}$

15. $g(x) = \begin{cases} -4 & \text{if } x \leq -3 \\ -2 & \text{if } -3 < x \leq -1 \\ 1 & \text{if } -1 < x \leq 1 \\ 3 & \text{if } 1 < x \leq 3 \\ 5 & \text{if } x > 3 \end{cases}$

The Least You Need to Know

- Linear inequalities in two variables describe relations in which one variable is greater or less than an expression involving the other variable.

- Graph inequalities as though they were equations, and shade the half-plane above or below the line to show whether y is greater or less than the expression in x. Use a dotted line if y is strictly greater or strictly less, or a solid line if y may also be equal to the expression.

- Fit a linear equation to a set of data using a graphing utility or by finding the equation of the line that passes through the point $(\bar{x}, \bar{y})$ and has a slope $m = \frac{(\text{sum of the } x\text{'s})(\text{sum of the } y\text{'s}) - (\text{number of points})(\text{sum of the } xy\text{'s})}{(\text{sum of the } x\text{'s})^2 - (\text{number of points})(\text{sum of the } x^2\text{'s})}$. The equation can be used to make predictions of the response variable (y) for other values of x.

- To evaluate a piecewise function, first determine in which section of the domain your input belongs and which rule applies to that section. You then evaluate using that rule and only that rule. A function produces only one y for any x.

- Graph a piecewise function by dividing the coordinate plane into sections to match the function rule.

Absolute Value

When you first meet the idea of the absolute of a number, you're usually talking about a distance, and the direction isn't important, so the sign can be dismissed. When you're asked in algebra to solve equations or inequalities that involve absolute values, however, you have to consider carefully what signs the numbers in the absolute value signs might have had. In this chapter, you learn to do that careful logic for equations, for simple inequalities, and for problems that combine two inequalities with a conjunction. You also consider absolute value as a type of function, and explore how transformations change its parent graph.

In This Chapter

- Solving absolute value equations
- Using compound inequalities to determine solutions of absolute value inequalities
- Graphing absolute value functions by transformations

Absolute Value Equations

That willingness to ignore direction is the primary complication in solving an *absolute value* equation. If you write $x = 4$, anyone who reads that knows that the value of x is 4. If you write $|x| = 4$, however, you know that x may be 4 or x may be -4. Every time you solve an equation involving an absolute value, you have to consider the two possibilities. The variable or expression inside the absolute value signs may be a positive number or a negative number.

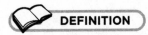

DEFINITION

The **absolute value** of a number is its distance from 0, regardless of direction—in other words, the size or magnitude of a number without regard to its sign. Formally, the absolute value function is defined as a piecewise function:

$$|x| = \begin{cases} x & \text{if } x \geq 0 \\ -x & \text{if } x < 0 \end{cases}$$

To solve an equation involving an absolute value, focus first on getting that absolute value isolated on one side of the equation. If you need to solve $3 + 2|4x - 1| = 77$, for example, you'll want to first find out what $|4x - 1|$ is worth. To do that, subtract 3 from both sides, and divide both sides by 2.

$$3 + 2|4x - 1| = 77$$
$$2|4x - 1| = 74$$
$$|4x - 1| = 37$$

Once you've isolated the absolute value, it's time to consider the two cases. If $|4x - 1| = 37$, the expression inside the absolute value signs, $4x - 1$, may be 37 or −37. Set up two equations to let you explore both possibilities.

$$|4x - 1| = 37$$
$$4x - 1 = 37 \quad or \quad 4x - 1 = -37$$
$$4x = 38 \quad or \quad 4x = -36$$
$$x = 9.5 \quad or \quad x = -9$$

However, an absolute value equation might only have one solution, if you find that the absolute value of some expression is equal to 0. Zero is unique in that it is its own opposite, so both cases are identical. There are also absolute value equations that have no solution—equations in which you find that the absolute value is equal to a negative number. Because the absolute value of any number—negative, 0, or positive—is considered to be positive, an equation like $|x - 5| = -9$ has no solution. There is no number that has an absolute value of −9.

ALGEBRA TRAP

Absolute value equations generally produce two solutions, but don't assume that those solutions are a number and its opposite. Isolate the absolute value, break it into a positive and a negative case, but then follow each case through algebraically to its solution. They may be very different numbers.

CHECKPOINT

Solve each equation.

1. $|4x - 3| = 9$

2. $|5 - 3x| + 7 = 25$

3. $|8 - 3x| - 5 = 12$

4. $4 - 2|x + 1| = -18$

5. $7|8x - 3| + 12 = 89$

Absolute Value Inequalities

When an absolute value appears in an inequality, you again need to consider two cases, but a thoughtful approach is key. Solutions of linear inequalities include not just a single number, but all the numbers above or below that number. Solving an inequality means doing that twice, with a sign change involved, and sign changes flip inequalities. So let's think this through bit by bit.

How Absolute Value Inequalities Work

If you only focus on the positive side of the number line, the inequality $|x| \geq 5$ is solved by the number 5 and all the numbers above 5, or to the right of 5 on the number line. When you consider negative numbers, you can see that -5 will make the sentence true, because the absolute value of -5 is 5, but numbers greater than -5 don't work. For instance, $|-4|$ is less than 5. On the other hand, numbers less than -5 do make the sentence true. $|-6|$, for example, is greater than 5. When you flip over to the negative side, you also have to flip the inequality. So the solution of $|x| \geq 5$ includes all numbers $x \geq 5$ or numbers $x \leq -5$. If you look at this on a number line as in the following illustration, you'll see two rays or arrows pointing in opposite directions.

The inequality $|x| \leq 4$ also involves a flip. On the positive side, numbers from 0 up to 4 are part of the solution, but on the negative side, you want numbers from -4 up to 0, not numbers less than -4. Therefore, the solution of $|x| \leq 4$ becomes $-4 \leq x \leq 4$.

Based on what you've learned from these examples, the solutions for inequalities based on their signs can be defined as follows:

The inequality $|x| < c$ is equivalent to $-c < x < c$ or $-c < x$ and $x < c$.

The inequality $|x| > c$ is equivalent to $x < -c$ or $x > c$.

Solving Absolute Value Inequalities

To solve an absolute value inequality, first isolate the absolute value, and then translate the absolute value inequality into the appropriate *compound inequality*. Finally, solve each of the cases in the compound inequality. Let's go through a couple examples to illustrate this.

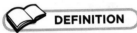

DEFINITION

A **compound inequality** is made up of two or more inequalities connected by a conjunction, such as *and* or *or*. For example, the inequality "$x > 3$ or $x < -2$" is a compound inequality using *or*, and the inequality "$-4 < x < 5$" is the condensed form of the compound inequality "$x > -4$ and $x < 5$."

To solve $2|3x - 7| + 5 < 15$, first subtract 5 from both sides and divide by 2 to isolate the absolute value.

$$2|3x - 7| + 5 < 15$$
$$2|3x - 7| < 10$$
$$|3x - 7| < 5$$

Now, because this says the absolute value is less than 5, convert it to $-5 < 3x - 7 < 5$ or $-5 < 3x - 7$ and $3x - 7 < 5$, and then solve.

$$-5 < 3x - 7 \quad \text{and} \quad 3x - 7 < 5$$
$$2 < 3x \quad \text{and} \quad 3x < 12$$
$$\tfrac{2}{3} < x \quad \text{and} \quad x < 4$$

The solution of $2|3x - 7| + 5 < 15$ is $\tfrac{2}{3} < x < 4$.

Let's try another example and solve $1 - |3 - 2x| \le -10$. To start, isolate the absolute value by subtracting 1 and dividing both sides by -1. Remember that when you divide both sides of an inequality by a negative number, the inequality reverses direction.

$$1 - |3 - 2x| \le -10$$
$$-|3 - 2x| \le -11$$
$$|3 - 2x| \ge 11$$

Convert the greater than inequality into the compound inequality $3 - 2x \le 11$ or $3 - 2x \ge 11$ and solve.

$$3 - 2x \leq -11 \quad \text{or} \quad 3 - 2x \geq 11$$
$$-2x \leq -14 \quad \text{or} \quad -2x \geq 8$$
$$x \geq 7 \quad \text{or} \quad x \leq -4$$

The solution of $1 - |3 - 2x| \leq -10$ is $x \leq -4$ or $x \geq 7$.

ALGEBRA TIP

To remember the correct conjunction when converting an absolute value inequality to a compound inequality, remember "Greator, less thand." If the absolute value is greater than a number, use *or*; if it's less than a number, use *and*.

CHECKPOINT

Solve each inequality.

6. $|x + 7| > 3$

7. $|2x + 5| - 7 > 22$

8. $3 - |3 - 5x| \leq 1$

9. $4 + |2x + 3| \geq 11$

10. $|9 - 2x| + 7 < 20$

Absolute Value Functions

The absolute value function commonly written as $f(x) = |x|$ is actually defined as a piecewise function: $f(x) = \begin{cases} x & \text{if } x \geq 0 \\ -x & \text{if } x < 0 \end{cases}$. That means that the graph of $f(x) = |x|$ is made up of the portion of $y = x$ to the right of the y-axis and the portion of $y = -x$ to the left of the y-axis. This gives the graph its characteristic V shape. The V has its *vertex* at the origin, the right side has a slope of 1, and the left side a slope of -1. This graph, shown as follows, is called the *parent function* for the absolute value family.

DEFINITION

The **vertex** of an absolute value graph is the point of the V. In the parent graph, the vertex is at the origin.

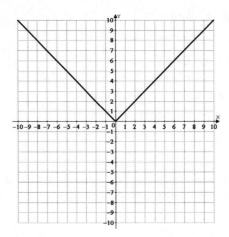

You can sketch other graphs in the absolute value family quickly by transforming this parent graph. When you transform a parent graph, it's a good idea to focus on a few key points that define the shape of the graph. In the case of the absolute value parent, the convenient choices are (0, 0), (1, 1), and (–1, 1).

Changes made inside the absolute value signs cause changes to the graph in a horizontal direction—a slide to the left or the right, a reflection across the y-axis, or a stretch or compression in the horizontal direction. Changes like these that affect the x-coordinates seem to do the opposite of what they say. For instance, subtracting a number from x moves the graph to the right, but adding moves the graph left. The following table provides examples of how different changes inside the absolute value signs can affect the x-coordinates of the parent graph.

Equation	Graph	(0, 0), (1, 1), (–1, 1)
$y = |x - 3|$		(3, 0), (4, 1), (2, 1)
$y = |x + 4|$		(–4, 0), (–3, 1), (–5, 1)

Equation	Graph	(0, 0), (1, 1), (−1, 1)
$y = \lvert -x \rvert$		(0, 0), (−1, 1), (1, 1)
$y = \lvert 2x \rvert$		$(0, 0), (\frac{1}{2}, 1), (-\frac{1}{2}, 1)$
$y = \left\lvert \frac{1}{3}x \right\rvert$		(0, 0), (3, 1), (−3, 1)

Changes to the equation outside the absolute value signs affect the y-coordinates, and so move the graph up or down, reflect it over the x-axis, or stretch or compress vertically. Unlike the changes inside the absolute value, these changes do what they seem to say. The following table provides examples of how different changes outside the absolute value signs can affect the y-coordinates of the parent graph.

Equation	Graph	(0, 0), (1, 1), (–1, 1)
$y = \lvert x \rvert - 3$		(0, –3), (1, –2), (–1, –2)
$y = \lvert x \rvert + 4$		(0, 4), (1, 5), (–1, 5)
$y = -\lvert x \rvert$		(0, 0), (1, –1), (–1, –1)
$y = 2\lvert x \rvert$		(0, 0), (1, 2), (–1, 2)
$y = \frac{1}{3}\lvert x \rvert$		(0, 0), (1, $\frac{1}{3}$), (–1, $\frac{1}{3}$)

When sketching the graph of an absolute value function that involves more than one of these transformations, you apply horizontal transformations first, starting with translations, followed by stretches or compressions, and reflection last. For the vertical transformations, the order is reversed: reflection, and then stretch or compress, and finally translations.

For example, to graph $f(x) = 2|x - 4| + 1$, first shift the parent graph 4 units to the right.

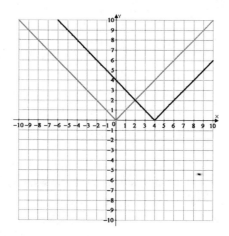

Stretch the graph vertically by a factor of 2.

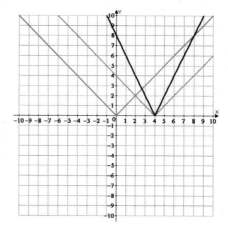

Finally, shift the graph up 1 unit.

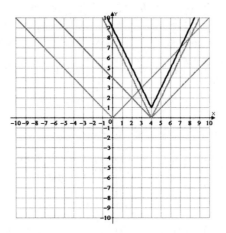

Now let's look at the graph of $g(x) = 4 - |1 - x|$. This will be a little easier if you rewrite the equation as $g(x) = -|-x + 1| + 4$. This allows you to see the sequence of transformations better—move 1 to the left, reflect over the y-axis, reflect over the x-axis, and finally slide up 4. The following chart shows the result.

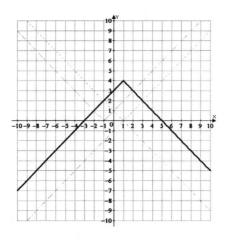

CHECKPOINT

Use transformations to sketch the graph of each function.

11. $f(x) = 2|x - 4|$

12. $g(x) = |x + 5| - 3$

13. $y = 3 - |x - 2|$

14. $f(x) = 5 - |x + 1|$

15. $g(x) = 3|x - 7| + 5$

The Least You Need to Know

- Each absolute value equation has two solutions.

- Isolate the absolute value and then consider whether the expression in the absolute value signs may be a positive or a negative number. Each possibility gives one solution.

- Absolute value inequalities convert to compound inequalities. When the absolute value is isolated, change |expression| > c to expression < $-c$ or expression > c or change |expression| < c to $-c$ < expression < c.

- The graph of the absolute value function is a V-shaped graph. Transform the parent graph using the key points (1, 1), (0, 0), and (–1, 1).

Solving Systems Algebraically

A thorough understanding of linear equations and functions is critical for successful work in algebra at any level. Combining linear relationships can take several forms. One, as you saw in the previous chapter, is the creation of absolute value functions. Another important method of combining linear equations is the creation of a system of equations. In algebra II, you expand your understanding of systems in several significant ways, which I go over in this chapter.

Systems of Equations

A *system of equations* in two variables is a pair of equations, each containing the same two variables. The equations $5x - 3y = -4$ and $2x + 9y = 29$, for example, form a system. Each equation in the pair has infinitely many possible solutions, but in many cases, it is possible to find a single pair of values that simultaneously solves both equations. Some systems have no solution, because the two equations are contradictory, and other systems have infinitely many solutions, because one equation is a constant multiple of the other.

In This Chapter

- Solving two-variable systems by graphing, substitution, and elimination
- Solving a three-variable system by substitution or elimination
- Showing the solution set of a system of inequalities by graphing
- Solving linear programming problems to find an optimal solution

DEFINITION

A **system of equations** is a collection of two (or more) equations in two (or more) variables. A unique solution requires as many equations as variables.

In algebra I, you learn three methods for solving systems of equations: graphing, substitution, and elimination. As you expand your understanding of systems in algebra II, you'll revisit and expand upon those methods, so a quick review is in order.

Graphing

One way a system of equations in two variables can be solved is by graphing each equation. Each graph is made up of a set of points that represent solutions of the equation, and so the point at which they overlap—the point of intersection—represents the pair of values that solve both equations, otherwise known as the solution of the system.

To solve the following system, for example, graph each equation by any convenient method.

$$5x - 3y = -4$$
$$2x + 9y = 29$$

The graph is shown in the following illustration.

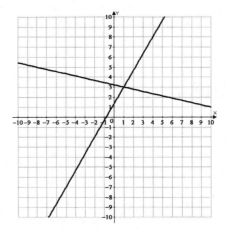

The point at which the two lines intersect, (1, 3), represents the solution of the system: $x = 1$, $y = 3$.

Two lines can cross at one point, not cross at all, or completely coincide. That's why a system of two linear equations has one solution, no solution, or infinitely many solutions. Later on, when you solve nonlinear systems, you'll find the graphs may cross more than once, giving you many more possibilities for the number of solutions. (You'll see more of that in Chapter 9.)

As a solution method, however, graphing has limitations, most significantly the fact that any solution other than a pair of integers is difficult to read from the graph. Even with a grid scale fine enough to show tenths, you may still have to approximate solutions. Substitution and elimination are, therefore, more practical than graphing.

CALCULATOR CORNER

Most graphing utilities—calculators or software packages—have an option to find the point(s) of intersection of two graphs. You'll need to select the two graphs you're interested in and perhaps move your cursor near the intersection point, and the software will then give you the coordinates of the point of intersection. To use this feature most effectively, zoom in or set a window that allows you to see the intersection clearly.

CHECKPOINT

Solve each system of equations by graphing.

1. $y - 2x = 8$
   $3x + 5y = 14$

2. $2y = 5x - 6$
   $3x + 16 = 4y$

3. $2x - 3y = 9$
   $x - 2y = 8$

4. $y - 6 = 3x + 1$
   $4x - 3y = 4$

5. $y = -\frac{2}{3}x - 2$
   $4x - 3y = 6$

Substitution

Solving a system by the *substitution method* requires transforming one of the equations so a variable is isolated. This can sometimes result in an expression that involves a lot of fractions, and that can discourage the use of substitution. In the system from the previous example, isolating a variable results in fractions no matter which equation or variable you choose.

Equation	Solved for x	Solved for y
$5x - 3y = -4$	$x = \frac{3y-4}{5}$	$y = \frac{5x+4}{3}$
$2x + 9y = 29$	$x = \frac{29-9y}{2}$	$y = \frac{29-2x}{9}$

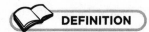

DEFINITION

The **substitution method** is a method of solving a system of equations by isolating one variable in one equation, and replacing that variable in the other equation with the expression it equals. The equation is then solved for the remaining variable, and that value is substituted back to find the second variable.

You can still substitute, of course, but the resulting equation will need a good bit of simplifying.

However, when isolating a variable in one of the equations has a friendlier result, substitution can be a convenient method of solution. In the following system, for example, you can isolate y in the second equation easily.

$$5x - 3y = -5$$
$$2x + y = 9$$

Use $y = 9 - 2x$ to substitute for y in the first equation. It becomes $5x - 3(9 - 2x) = -5$, which can be solved to find the value of x.

$$5x - 3(9 - 2x) = -5$$
$$5x - 27 + 6x = -5$$
$$11x - 27 = -5$$
$$11x = 22$$
$$x = 2$$

Once the value of x is known, you can substitute that value, 2 in this case, into either of the original equations to solve for y.

$$2x + y = 9$$
$$2(2) + y = 9$$
$$4 + y = 9$$
$$y = 5$$

The solution of the system is $x = 2$, $y = 5$.

CHECKPOINT

Solve each system of equations by substitution.

6. $y = 2x - 7$
    $3x + y = 18$

7. $x + y = 8$
    $3x - 7y = -6$

8. $x - 3y = -7$
    $2y - 3x = 14$

9. $6x - y = 17$
    $5x + 2y = 34$

10. $4x - 2y = 42$
    $3x + 5y = 12$

Elimination

The *elimination method* of solving a system is based on the fundamental idea of adding the same amount to both sides of the equation, or the addition property of equality. You start with the top equation, and then add the bottom equation to it, left side to left side and right side to right side. The two sides are equal, so you're adding the same amount to both sides.

In a simple example, like the following system, adding the two equations eliminates one variable, making it easy to solve for the remaining variable and plug that back in to find the other.

$$x + y = 23$$
$$\underline{x - y = -1}$$
$$2x = 22$$
$$x = 11$$
$$11 + y = 23$$
$$y = 12$$

The solution of the system is $x = 11$ and $y = 12$.

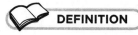

DEFINITION

The **elimination method** for solving a system adds the two equations, or multiples of the equations, so the resulting equation contains only one variable. The value of that variable is found and substituted back in to find the other variable.

To solve the system from the previous section by elimination, you will need to multiply the bottom equation by 3 first to make the coefficients of the y-terms opposites.

$$5x - 3y = -5$$
$$\underline{3(2x + y) = (9) \cdot 3} \qquad\qquad 2x + y = 9$$
$$5x - 3y = -5 \qquad\qquad\quad 2(2) + y = 9$$
$$\underline{6x + 3y = 27} \qquad\qquad\quad 4 + y = 9$$
$$11x = 22 \qquad\qquad\qquad\quad y = 5$$
$$x = 2$$

The solution of the system is $x = 2$ and $y = 5$.

CHECKPOINT

Solve each system by elimination.

11. $7x - 3y = 1$
     $2x - y = 1$

12. $2x + 3y = 21$
     $4x + y = 9$

13. $3x - 7y = 30$
     $5x + y = 12$

14. $2x + 4y = 5$
     $4x + 5y = 7$

15. $2x + 3y = 4$
     $3x - 8y = -9$

Systems with Three Variables

The first expansion of systems you meet in algebra II is solving systems that involve three variables, rather than just two. Later, you'll see that there are methods for solving systems with any number of variables, but for now, three will be enough. Each of the methods of solving a system with three variables is an adaptation of the methods you used for two variables, with the exception of graphing.

Theoretically, while you could solve a three-variable system by graphing each equation, three-dimensional graphing is a challenge for those of us with less than stellar artistic skills, and so is not useful for systems. However, substitution and elimination can be used.

Using Substitution

Three-variable systems can be solved by substitution, with the same limitation as two-variable systems: the expressions may become uncomfortably complicated. Consider the following system, for example, which certainly could be solved by substitution.

$$x + y + z = 8$$
$$x - y + z = -2$$
$$x + y - z = 16$$

Choose one equation and isolate a variable. Here, let's isolate x in the first equation: $x = 8 - y - z$. In each of the other two equations, replace x with $8 - y - z$.

$$(8 - y - z) - y + z = -2$$
$$(8 - y - z) + y - z = 16$$

Simplify each of these equations.

$$(8 - y - z) - y + z = -2$$
$$8 - 2y = -2$$
$$-2y = -10$$
$$y = 5$$
$$(8 - y - z) + y - z = 16$$
$$8 - 2z = 16$$
$$-2z = 8$$
$$z = -4$$

Finally, go back and find the value of x.

$$x + y + z = 8$$
$$x + 5 - 4 = 8$$
$$x + 1 = 8$$
$$x = 7$$

The solution of the system is $x = 7$, $y = 5$, and $z = -4$.

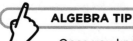

ALGEBRA TIP

Once you know the value of one of the variables, make sure you substitute it back into the original version of one of the equations to find the other variables. Don't use any transformed version of them; they may have errors.

Of course, not every system is that quick to solve. Let's try substitution on one a little more complicated.

$$x - 2y + 20z = 1$$
$$3x + y - 4z = 2$$
$$2x + y - 8z = 3$$

To avoid fractions, solve for x in the first equation or for y in the second or third equation. In this case, let's isolate y in the bottom equation.

$$2x + y - 8z = 3$$
$$y = 3 - 2x + 8z$$

Replace y in each of the first two equations with $3 - 2x + 8z$.

$$x - 2(3 - 2x + 8z) + 20z = 1$$
$$3x + (3 - 2x + 8z) - 4z = 2$$

Simplify each equation.

$$x - 2(3 - 2x + 8z) + 20z = 1 \qquad 3x + (3 - 2x + 8z) - 4z = 2$$
$$x - 6 + 4x - 16z + 20z = 1 \qquad 3x - 2x + 8z - 4z + 3 = 2$$
$$5x + 4z - 6 = 1 \qquad x + 4z + 3 = 2$$
$$5x + 4z = 7 \qquad x + 4z = -1$$

Now you have a two-variable system to solve, and you can choose to solve it by substitution or elimination. Let's go with substitution by isolating x in the second equation: $x + 4z = -1$ becomes $x = -1 - 4z$. Substitute and solve.

$$5x + 4z = 7$$
$$5(-1 - 4z) + 4z = 7$$
$$-5 - 20z + 4z = 7$$
$$-5 - 16z = 7$$
$$-16z = 12$$
$$z = -\frac{12}{16} = -\frac{3}{4}$$

 **ALGEBRA TRAP**

With all the work that goes into solving a three-variable system, it's easy to get to the value of one variable, give a sigh of relief, and move on to the next battle. Remember, you're not done until you have found the values of all the variables. Don't stop before you reach that goal.

Back up to one of the two-variable equations and replace z with $-\frac{3}{4}$ to find the value of x. Using $x + 4z = -1$, you get $x + 4\left(-\frac{3}{4}\right) = -1$ or $x - 3 = -1$, so $x = 2$. Don't forget to go back to find y.

$$2x + y - 8z = 3$$
$$2(2) + y - 8\left(-\frac{3}{4}\right) = 3$$
$$4 + y + 6 = 3$$
$$y + 10 = 3$$
$$y = -7$$

Using Elimination

The elimination method for three-variable systems is a double elimination. The first phase eliminates one variable, leaving you a system of two equations with two variables. The second phase is solving that system, and of course, plugging back in to find the value of the third variable.

When you solve a three-variable system like the following, your first task is to decide which variable you want to eliminate.

$$3x + y - 4z = 2$$
$$x - 2y + 20z = 1$$
$$2x + y - 8z = 3$$

It's your choice, so pick whichever variable you think you can eliminate most easily. However, choose before you begin, so you keep your focus on that task and don't get distracted. In this case, let's eliminate y. To do this, make two pairs of equations. You can do first and second, first and third, or second and third. Again, it's your choice. For this example, let's use the first and second for one pair and the second and third for the other pair.

$$3x + y - 4z = 2 \qquad x - 2y + 20z = 1$$
$$x - 2y + 20z = 1 \qquad 2x + y - 8z = 3$$

Keep your focus on eliminating y's in each pair. You'll need to do some multiplying first.

$$2(3x + y - 4z) = (2) \cdot 2 \qquad x - 2y + 20z = 1$$
$$\underline{x - 2y + 20z = 1} \qquad \underline{2(2x + y - 8z) = (3) \cdot 2}$$
$$6x + 2y - 8z = 4 \qquad x - 2y + 20z = 1$$
$$\underline{x - 2y + 20z = 1} \qquad \underline{4x + 2y - 16z = 6}$$
$$7x + 12z = 5 \qquad 5x + 4z = 7$$

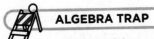

ALGEBRA TRAP

When solving a three-variable system, it's crucial that you have a plan before you begin. If you just go in looking for the easiest thing to do, you may find yourself eliminating one variable here, another there, and never getting down to a solution. Don't play a game of algebraic whack-a-mole. Know which variable you're going to eliminate, and keep your focus on that until you achieve the goal.

Now you have two equations containing x and z. You need to solve this system, and you can choose the method you think is most convenient. Let's use elimination here, too. Multiply $5x + 4z = 7$ by -3 and add the equations to eliminate z.

$$7x + 12z = 5$$
$$-3(5x + 4z) = (7)(-3)$$
$$\overline{}$$
$$7x + 12z = 5$$
$$-15x - 12z = -21$$
$$\overline{}$$
$$-8x = -16$$
$$x = 2$$

Replace x with 2 in one of the two-variable equations to find z. $5x + 4z = 7$ becomes $10 + 4z = 7$, so $z = -\frac{3}{4}$. Return to one of the original equations and solve for y.

$$3x + y - 4z = 2$$
$$3(2) + y - 4\left(-\tfrac{3}{4}\right) = 2$$
$$6 + y + 3 = 2$$
$$y + 9 = 2$$
$$y = -7$$

The solution of this system is $x = 2, y = -7, z = -\frac{3}{4}$.

CHECKPOINT

Solve each system by substitution, elimination, or a combination of the two.

16. $2x - y - 3z = 10$
    $x - 2y + 3z = -22$
    $3x + 5y - z = 63$

17. $4x - 2y + 3z = -20$
    $2x + 2y - z = 28$
    $x - 2y + 3z = -32$

18. $3x - y + 4z = 21$
    $5x + 2y - 3z = -6$
    $5x - 6y + z = -2$

19. $x - 7y - 3z = 6$
    $3x + y - z = 8$
    $x - y + 5z = 12$

20. $x + y - z = -6$
    $x - y + z = 14$
    $x - y - z = 8$

Systems of Inequalities

You've probably noticed that whenever you talk about linear equations, you soon find yourself talking about linear inequalities. Systems are no exception. There are whole classes of problems that depend on systems of linear inequalities.

Graphing is the preferred method for discovering the solution of a *system of inequalities*. Remember that inequalities are relations, not functions, with each *x*-value corresponding to many different *y*-values, which makes describing the solution of even one inequality in words or symbols a difficult task. Showing the collection of points that represent the solution set via a graph is the simpler course of action.

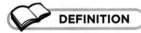

DEFINITION

A **system of inequalities** is a set of two (or more) inequalities in two (or more) variables, which together define a region that contain all the points that simultaneously solve all the inequalities.

Solving a system of inequalities simply means carefully graphing each inequality, including its shading, on one set of axes. Although you won't get the satisfaction of seeing that single point of intersection that appears when you solve a system of equations by graphing, you will usually see the shading overlap. That overlap is the solution set of the system of inequalities.

To solve the following system of inequalities, start by carefully graphing the first inequality.

$$4x - y < -9$$
$$3x + 2y \geq -4$$

You can use any convenient method to draw the graph. For the first inequality, let's isolate y and use slope and y-intercept to draw the graph: $4x - y < -9$ becomes $4x + 9 < y$ or $y > 4x + 9$. The graph has a y-intercept of 9 and a slope of 4. You'll graph it with a dotted line, because the inequality does not include an "or equal to," and the shading will go up, because you want the points where y is greater than $4x + 9$.

The second inequality, $3x + 2y \geq -4$, can be transformed to isolate y, and will become $y \geq -\frac{3}{2}x - 2$, so it will have a y-intercept of -2 and a slope of $-\frac{3}{2}$. Draw it with a solid line and shade up. Here's what the graph of each inequality would look like if you graphed them separately.

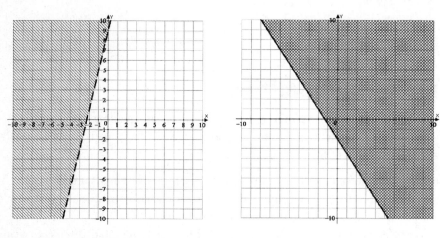

ALGEBRA TIP

It is possible to have a system of inequalities with no solution. This happens when the lines are parallel and the shaded regions do not overlap.

For a system of inequalities, however, you want to graph both of these on the same set of axes. The following figure shows what that will look like. The wedge-shaped region shaded by both inequalities, bordered by a solid line on the left and a dotted line on the right, is the solution set of the system.

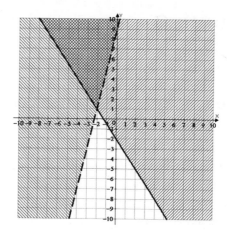

 ALGEBRA TRAP

Don't forget that multiplying or dividing both sides of an inequality by a negative number will reverse the direction of the inequality sign. If you miss that detail, the shading will be on the wrong side of the line, and you'll wind up with the wrong shaded region.

 CHECKPOINT

Graph each system of inequalities. Mark the solution set clearly.

21. $y - 2x \leq 8$
    $3x + 5y > 14$

22. $2y < 5x - 6$
    $3x + 16 \leq 4y$

23. $2x - 3y > 9$
    $x - 2y < 8$

24. $y - 6 \leq 3x + 1$
    $4x - 3y \geq 4$

25. $y > -\frac{2}{3}x - 2$
    $4x - 3y \leq 6$

Linear Programming

In systems of inequalities, especially when the number of equations exceeds the number of variables, the solution is a region containing infinitely many points, each of which satisfies the conditions specified by all the inequalities. When those inequalities represent real limitations on a real-life situation, it's nice to know you have lots of possible solutions. But which one do you pick? *Linear programming* is a process of finding the best solution when there are many possible solutions available.

Consider the situation of a young entrepreneur manufacturing carry bags for wheelchair-bound adults. She has designed bags in two styles: a briefcase type and a day pack model. Each style of bag will consume some of her resources, whether material, worker time, or budget. She needs to determine how she will allocate those resources—that is, how many briefcases and how many day packs she will produce.

She can define her variables as x = the number of briefcases to produce and y = the number of day packs. She has a limited number of hours of work available each month. The total production must be held to a level that uses no more of any material than she can acquire in a month. If she has a contract to fulfill, that will put a lower limit on what she must produce.

ALGEBRA TIP

When you are graphing several inequalities on the same grid, shading the entire half-plane for every inequality can obscure details of the graph. Instead, just put a narrow fringe along the side of the line that you plan to shade. When you've done that for all the inequalities, the region that would have multiple shadings will have fringe on all sides.

Let's put some numbers on this situation. A skilled worker can assemble a day pack in 3 hours, but the more detailed briefcase will require 5 hours to construct. There are 480 hours of labor available each month. The briefcase is closed with 1 fastener and the day pack with 2. These fasteners are purchased from a supplier, and in a given month, 120 fasteners are available. A contract with a retailer requires that the shipment of merchandise must contain at least 50 pieces—that is, the number of briefcases and the number of day packs, in whatever combination, must total at least 50.

Producing x briefcases will consume $5x$ hours of worker time and y day packs will use $3y$ hours. The total time cannot exceed 480 hours, so $5x + 3y \leq 480$. Producing x briefcases will use x fasteners, while y day packs will use $2y$ fasteners; therefore, $x + 2y \leq 120$. Her contract requires that $x + y \geq 50$.

Finally, in most of the situations linear programming explores, the variables represent numbers of physical objects or units of time; therefore, they must be non-negative and in many cases, they must be whole number values. In the example, because both x and y are numbers of bags, it makes sense that $x \geq 0$ and $y \geq 0$. Usually, this just means that you're working exclusively in the first quadrant.

Together these inequalities form a set or system of *constraints,* which you can graph as follows.

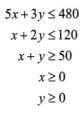

$$5x + 3y \leq 480$$
$$x + 2y \leq 120$$
$$x + y \geq 50$$
$$x \geq 0$$
$$y \geq 0$$

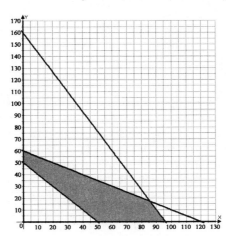

If you graph the system of inequalities, the solution of the system—the area shaded by all the inequalities—represents what is called the *feasible region,* the set of all points that could work. Any point in the feasible region represents a number of briefcases and a number of day packs that the entrepreneur could produce while staying within the constraints. But which one should she use? Which is the best solution?

What's best will vary from situation to situation, depending on your goal. Do you want to keep costs down, make the maximum profit, or something else? You define what *best* means by an *objective function*—an equation like Cost = $12x + 9y$ or Profit = $8x + 10y$ that you want to maximize or minimize.

Let's suppose the business person in the example makes $12 profit on each briefcase and $9 profit on each day pack and wants to maximize her profit. The best, or *optimal,* solution will occur at a vertex, or corner, of the feasible region. Identify each vertex and evaluate the objection function for the values of x and y at each vertex. The feasible region has vertices shown in the following table. (I rounded to the nearest whole number, because a fraction of a bag or briefcase wouldn't be profitable.)

x	y	Profit = $12x + 9y$
0	50	$12(0) + 9(50)= 450$
0	60	$12(0) + 9(60)= 540$
86	17	$12(86) + 9(17)=1{,}185$
50	0	$12(50) + 9(0)= 600$
96	0	$12(96) + 9(0)= 1{,}152$

DEFINITION

Linear programming is a technique for finding the best, or optimal, solution in a situation where choices are limited by a set of constraints but many solutions are possible within those constraints. **Constraints** are limitations on the activity, represented by inequalities. The **objective function** is an equation that defines what you want to maximize or minimize in choosing the best solution. The **feasible region** is the area that is shaded by all the inequalities that represent the constraints. It's called the *feasible region* because each point is a feasible, or workable, solution within the constraints. The **optimal solution** is the point within the feasible region that maximizes or minimizes the objective function.

The combination of 86 briefcases and 17 day packs will give the maximum profit while staying within all the necessary limitations.

CHECKPOINT

Determine the vertex of the feasible region that represents the maximum value of the objective function.

26. Profit = $14x + 21y$ subject to $4x + 7y \leq 29$, $7x + 3y \leq 23$, $x \geq 0$, and $y \geq 0$.

27. Profit = $7.25x + 3.75y$ subject to $9x + 4y \leq 37$, $5x + 8y \leq 35$, $x \geq 0$, and $y \geq 0$.

Determine the vertex of the feasible region that represents the minimum value of the objective function.

28. Cost = $20x + 30y$ subject to $10x + 3y \geq 30$, $4x + 6y \geq 28$, $0 \leq x \leq 5$, and $0 \leq y \leq 9$.

29. Cost = $150x + 120y$ subject to $x - y \geq -3$, $5x - 2y \leq 6$, and $x + 3y \geq 6$.

30. A small bakery sells cookies and muffins, and bakes large batches of each. Each batch of cookies uses 2 pounds of flour, 3 cups of sugar, and 1 pound of butter. Each batch of muffins uses 8 pounds of flour, 5 cups of sugar, and 1 pound of butter. On any day, there are 52 pounds of flour, 36 cups of sugar, and 10 pounds of butter. The bakery must make at least 1 batch of cookies and at least 1 batch of muffins. The bakery estimates a profit of $12 on each batch of cookies and $15 on each batch of muffins. How many batches of each should the bakery make to maximize profit?

The Least You Need to Know

- Solve a system of two-variable equations by graphing and finding the point of intersection, by isolating one variable and substituting into the other equation, or by adding the equations or multiples of the equations in order to eliminate a variable.

- Three-variable systems can be solved by using substitution or elimination to reduce the system to two equations in two variables, and then solving as a two-variable system.

- Show the solution of a system of inequalities by graphing all the inequalities on the same set of axes and finding the region that is part of all the shaded areas.

- Linear programming problems are made up of a set of inequalities that express the constraints on an activity and an equation, called the *objective function*, that defines the valued quantity to be maximized or minimized. The graphs of those inequalities define a feasible region, and the optimal solution occurs at a vertex of that region.

Systems and Matrices

Algebra is great at reducing a problem to the essentials. Whether you buy five DVDs and two games and spend a total of $100 or you have five shelves and two tables to display 100 items for sale, for example, algebra brings it down to the equation $5x + 2y = 100$. In this chapter, you learn to use an organizing device called a *matrix* to reduce a system of equations to its essentials and to give you new methods of solving that system. These methods may prove faster and easier, especially for large systems, and particularly if you have a calculator or computer to help with the matrices.

In This Chapter

- Operating with matrices
- Finding the determinant of a matrix
- Transforming matrices to solve systems
- Using an inverse matrix to solve a system

Matrices and Matrix Arithmetic

A *matrix* is just a rectangular arrangement of numbers (or other objects). Matrices use a structure of rows and columns to organize the numbers. The *dimension*, or order, of a matrix is simply the number of rows and columns (in that order) that the matrix contains. For instance, the dimension of the following matrix is 2-by-3.

	Column 1	Column 2	Column 3
Row 1	12	-4	19
Row 2	3	5	28

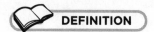

DEFINITION

A **matrix** is a device that displays numbers organized into rows and columns. The **dimension** of the matrix is the number of rows by the number of columns.

A system of linear equations can be reduced to a matrix of numbers by allowing each row to represent one equation, and each column a variable or the constants. The matrix contains just the coefficients and constants. Matrices can provide a way to perform repetitive arithmetic tasks by one matrix operation. Before you begin plugging information into a matrix, however, you need to know the rules of matrix arithmetic.

Addition and Subtraction

To add or subtract matrices, you must have matrices with exactly the same dimensions. So you can add a 2-by-3 matrix to a 2-by-3 matrix, but not to a 3-by-2 matrix. You need the exact same dimension because you add or subtract the matrices by adding or subtracting the numbers in corresponding positions.

Suppose you tracked new voter registrations each month in two states and broke the information down by party affiliation. You might organize that information into matrices like the following.

$$\begin{array}{cc} \text{January} & \text{February} \\ \begin{array}{ccc} \text{Rep} & \text{Dem} & \text{Ind} \end{array} & \begin{array}{ccc} \text{Rep} & \text{Dem} & \text{Ind} \end{array} \\ \begin{array}{c} \text{NY} \\ \text{NJ} \end{array} \begin{bmatrix} 293 & 276 & 82 \\ 187 & 193 & 14 \end{bmatrix} & \begin{array}{c} \text{NY} \\ \text{NJ} \end{array} \begin{bmatrix} 199 & 141 & 62 \\ 203 & 113 & 37 \end{bmatrix} \end{array}$$

If you want to know the change from month to month, you can subtract the matrices.

$$\begin{bmatrix} 293 & 276 & 82 \\ 187 & 193 & 14 \end{bmatrix} - \begin{bmatrix} 199 & 141 & 62 \\ 203 & 113 & 37 \end{bmatrix} = \begin{bmatrix} 293-199 & 276-141 & 82-62 \\ 187-203 & 193-113 & 14-37 \end{bmatrix} = \begin{bmatrix} 94 & 135 & 20 \\ -16 & 80 & -23 \end{bmatrix}$$

CALCULATOR CORNER

Most graphing calculators can perform matrix operations. In the matrix editing menu, you'll need to enter the dimension of the matrix, as well as the contents. Matrix arithmetic can be performed using the names of the matrices and the normal operations keys.

Multiplication

There are two types of multiplication in matrix arithmetic. Scalar multiplication multiplies a single number times a matrix. The single number, called a *scalar*, is multiplied by every number in the matrix, rather like the distributive property in regular arithmetic and algebra.

$$7\begin{bmatrix} 3 & 1 & -4 & 0 \\ 6 & -5 & 2 & -1 \end{bmatrix} = \begin{bmatrix} 7 \cdot 3 & 7 \cdot 1 & 7(-4) & 7 \cdot 0 \\ 7 \cdot 6 & 7(-5) & 7 \cdot 2 & 7(-1) \end{bmatrix} = \begin{bmatrix} 21 & 7 & -28 & 0 \\ 42 & -35 & 14 & -7 \end{bmatrix}$$

Multiplying a matrix by another matrix involves working with rows and columns. Let's look at one row and one column first to see how it's done. The row matrix comes first, followed by the column matrix; there must be as many numbers in the column as in the row, because you're going to pair them up. The following example has a row matrix with 3 numbers, so it's a 1-by-3, and a column matrix that has 3 numbers, making it a 3-by-1.

$$\begin{bmatrix} 3 & 5 & 2 \end{bmatrix} \begin{bmatrix} 7 \\ -2 \\ 4 \end{bmatrix}$$

A 1-by-3 times a 3-by-1 will produce a single number—a 1-by-1 matrix. Multiply the $3 \cdot 7$, the $5 \cdot -2$, and the $2 \cdot 4$, and add the products together to make that one number.

$$\begin{bmatrix} 3 & 5 & 2 \end{bmatrix} \begin{bmatrix} 7 \\ -2 \\ 4 \end{bmatrix} = \begin{bmatrix} 3(7) + 5(-2) + 2(4) \end{bmatrix} = \begin{bmatrix} 21 - 10 + 8 \end{bmatrix} = \begin{bmatrix} 19 \end{bmatrix}$$

To multiply larger matrices, the number of columns in the first matrix must equal the number of rows in the second. If you don't have that match, you can't multiply, so it's important to be sure you have the matrices in the right order. You do the actual multiplication by multiplying each row of the first matrix by each column of the second. Each row-times-column answer goes in the position that matches the row number and column number. So, for example, the product of row 3 and column 5 goes in the third row, fifth column of the answer.

 **CALCULATOR CORNER**

An attempt to multiply matrices for which the number of columns of the first does not equal the number of rows of the second will cause a dimension error. If this happens to you, check that your matrices are in the correct order.

In the following example, row 1 times each of the four columns looks like this.

$$\begin{bmatrix} 2 & -4 & 3 \\ -1 & 0 & 2 \end{bmatrix} \begin{bmatrix} 3 & 2 & 1 & 0 \\ 5 & -1 & -3 & 4 \\ -2 & -4 & 1 & 3 \end{bmatrix} = \begin{bmatrix} 6-20-6 & 4+4-12 & 2+12+3 & 0-16+9 \end{bmatrix} = \begin{bmatrix} -20 & -4 & 17 & -7 \end{bmatrix}$$

After those multiplications are done, multiply row 2 times each of the four columns.

$$\begin{bmatrix} 2 & -4 & 3 \\ -1 & 0 & 2 \end{bmatrix} \begin{bmatrix} 3 & 2 & 1 & 0 \\ 5 & -1 & -3 & 4 \\ -2 & -4 & 1 & 3 \end{bmatrix} = \begin{bmatrix} -20 & -4 & 17 & -7 \\ -3+0+-4 & -2+0-8 & -1+0+2 & 0+0+6 \end{bmatrix} = \begin{bmatrix} -20 & -4 & 17 & -7 \\ -7 & -10 & 1 & 6 \end{bmatrix}$$

As you can probably tell, matrix multiplication is not commutative. Normally in multiplication, the order doesn't matter; the product of 4 and 5 is 20 whether you do $4 \cdot 5$ or $5 \cdot 4$. If you switch the order of the matrices, however, the multiplication may not even be possible; if it is, you're unlikely to get the same answer.

CHECKPOINT

Given the matrices $A = \begin{bmatrix} 2 & -1 \\ -3 & 2 \end{bmatrix}$, $B = \begin{bmatrix} 3 & -2 \\ -1 & 2 \\ 1 & 0 \end{bmatrix}$, $C = \begin{bmatrix} 3 \\ -5 \end{bmatrix}$, $D = \begin{bmatrix} 1 & 2 & 0 \\ -4 & 3 & 5 \end{bmatrix}$,

$E = \begin{bmatrix} 2 & -3 & 1 \end{bmatrix}$, and $F = \begin{bmatrix} 1 & -2 \\ 3 & 1 \end{bmatrix}$, perform each calculation if possible.

1. $[A] + [F]$ 4. $[F] \cdot [D]$

2. $[B] - [D]$ 5. $[B] \cdot [C]$

3. $[E] \cdot [B]$

Determinants

If a matrix has the same number of rows as columns, it's called a *square matrix*. For every square matrix, there is a number called the *determinant*, derived from the elements of the matrix. If you name a square matrix with an uppercase letter, like $[A]$, you denote the determinant of that matrix as $|A|$.

DEFINITION

The **determinant** of a square matrix is a single number associated with a square matrix, found by operations with the numbers on the diagonals.

For a 2-by-2 matrix, finding the determinant requires just a little bit of arithmetic. The major diagonal of a square matrix runs from the upper left to the lower right, and the minor diagonal runs from the upper right to the lower left. The determinant of a 2-by-2 matrix is the product of the numbers on the major diagonal minus the product of the numbers on the minor diagonal. If $[A] = \begin{bmatrix} 1 & 2 \\ 3 & 4 \end{bmatrix}$, for instance, you find the determinant as follows.

$$|A| = \begin{vmatrix} 1 & 2 \\ 3 & 4 \end{vmatrix} = 1 \cdot 4 - 2 \cdot 3 = 4 - 6 = -2$$

For larger square matrices, the computation of the determinant becomes more complicated, and software or calculators equipped to handle matrices are invaluable. A 3-by-3 matrix is probably the largest determinant you would want to tackle by hand. To do that, duplicate the first two columns of the matrix on the right of the original.

For example, if $[A] = \begin{bmatrix} 1 & 2 & 3 \\ 4 & 5 & 6 \\ 7 & 8 & 9 \end{bmatrix}$, you start by duplicating the first two columns.

$$
\begin{array}{ccc|cc}
\boxed{1} & (2) & \langle 3 \rangle & 1 & 2 \\
4 & \boxed{5} & (6) & \langle 4 \rangle & 5 \\
7 & 8 & \boxed{9} & (7) & \langle 8 \rangle
\end{array}
$$

This allows you to find three diagonals that run upper left to lower right, marked in the previous example by boxes, parentheses, and pointed brackets. Find the product of each of these diagonals, and add them.

$$(1 \cdot 5 \cdot 9) + (2 \cdot 6 \cdot 7) + (3 \cdot 4 \cdot 8) = 45 + 84 + 96 = 225$$

Now look for three upper-right to lower-left diagonals, marked in a similar fashion as follows. Find and combine the products of those three.

$$
\begin{array}{ccc|cc}
1 & 2 & \boxed{3} & (1) & \langle 2 \rangle \\
4 & \boxed{5} & (6) & \langle 4 \rangle & 5 \\
\boxed{7} & (8) & \langle 9 \rangle & 7 & 8
\end{array}
\quad = 3 \cdot 5 \cdot 7 + 1 \cdot 6 \cdot 8 + 2 \cdot 4 \cdot 9 = 105 + 48 + 72 = 225
$$

CALCULATOR CORNER

Graphing calculators generally have a function for finding the determinant of a square matrix. Look for a function like det() in the matrix menu, and do not confuse the determinant with absolute value.

Finally, the determinant is the difference between the two calculations.

$$225 - 225 = 0$$

The determinant of the 3-by-3 matrix you started with is 0.

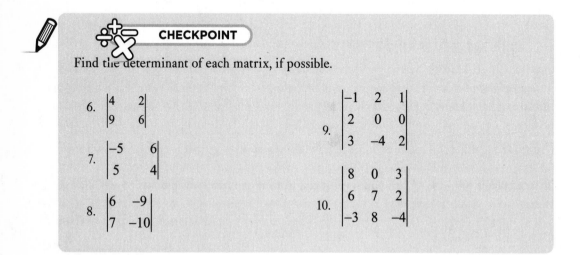

CHECKPOINT

Find the determinant of each matrix, if possible.

6. $\begin{vmatrix} 4 & 2 \\ 9 & 6 \end{vmatrix}$

7. $\begin{vmatrix} -5 & 6 \\ 5 & 4 \end{vmatrix}$

8. $\begin{vmatrix} 6 & -9 \\ 7 & -10 \end{vmatrix}$

9. $\begin{vmatrix} -1 & 2 & 1 \\ 2 & 0 & 0 \\ 3 & -4 & 2 \end{vmatrix}$

10. $\begin{vmatrix} 8 & 0 & 3 \\ 6 & 7 & 2 \\ -3 & 8 & -4 \end{vmatrix}$

Representing a System by Matrices

There are two ways that systems of linear equations are commonly represented by matrices. One creates a square matrix for the information about the variables and a column matrix to hold the constants.

Each row is an equation, and each column is a variable. So, for example, the system $5x - 3y = 11$ and $2x + 7y = 29$ translates into the two matrices shown as follows.

$$\begin{bmatrix} 5 & -3 \\ 2 & 7 \end{bmatrix} \qquad \begin{bmatrix} 11 \\ 29 \end{bmatrix}$$

The other method is to represent the whole system in one matrix, called an *augmented matrix*. The preceding system would be represented by the following augmented matrix.

DEFINITION

An **augmented matrix** is a rectangular matrix with R rows and C columns formed by appending an additional column to a square matrix.

$$\begin{bmatrix} 5 & -3 & | & 11 \\ 2 & 7 & | & 29 \end{bmatrix}$$

There are several matrix methods for solving systems. Some use the augmented matrix arrangement, and some use the square matrix and column matrix. Let's look at one that uses the square matrix and determinants.

Cramer's Rule

Cramer's rule, also known as the method of determinants, finds the solution of a system of equations based on the determinants of the matrix of coefficients and some variants on that matrix. The system $5x - 3y = 11$ and $2x + 7y = 29$ can serve as an example. The rule tells you to start with the determinant of the matrix of coefficients. Let's call that D.

$$D = \begin{vmatrix} 5 & -3 \\ 2 & 7 \end{vmatrix} = 5 \cdot 7 - 2(-3) = 35 + 6 = 41$$

Next, you create a variant of the coefficient matrix for each variable in the system. You replace the column that contains that variable's coefficient with the column of constants. For x, the matrix $\begin{bmatrix} 5 & -3 \\ 2 & 7 \end{bmatrix}$ becomes $\begin{bmatrix} 11 & -3 \\ 29 & 7 \end{bmatrix}$, and for y, $\begin{bmatrix} 5 & -3 \\ 2 & 7 \end{bmatrix}$ becomes $\begin{bmatrix} 5 & 11 \\ 2 & 29 \end{bmatrix}$. You now find the determinant of each of these new matrices. Let's call them N_x and N_y.

$$N_x = \begin{vmatrix} 11 & -3 \\ 29 & 7 \end{vmatrix} = 11 \cdot 7 - 29(-3) = 77 + 87 = 164 \quad \text{and} \quad N_y = \begin{vmatrix} 5 & 11 \\ 2 & 29 \end{vmatrix} = 5 \cdot 29 - 2 \cdot 11 = 145 - 22 = 123$$

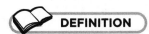 **DEFINITION**

> **Cramer's rule** is a method of solving a system of equations using the determinants of matrices derived from the system.

Finally, Cramer's rule tells you that $x = \frac{N_x}{D}$ and $y = \frac{N_y}{D}$. In this example, $x = \frac{N_x}{D} = \frac{164}{41} = 4$ and $y = \frac{N_y}{D} = \frac{123}{41} = 3$.

Cramer's rule can be applied to systems of any size, forming a variant matrix for each variable and finding the value of the variable by the determinant of its variant over the determinant of the coefficient matrix. You are only limited by your ability to find the determinants as the matrices become large.

CHECKPOINT

Solve by Cramer's rule.

11. $2x + y = 4$
$4x - 3y = 13$

12. $2x + 3y = -8$
$x + 2y = -3$

13. $3x + 7y = -10$
$-x + y = 2$

14. $5x + 10y \quad\quad = 70$
$5x \quad\quad + 25z = 270$
$10y + 25z = 300$

15. $2x - 3y + z = 16$
$x + 3y - 2z = -7$
$4x + 5y - 3z = -10$

CALCULATOR CORNER

On a graphing calculator that supports matrices, you can enter the matrix of coefficients, and then make copies and edit them to create the matrices for the numerators.

Solving Systems by Row Operations

There are transformations you can make to equations that produce equivalent equations—ones that change the appearance but produce equations with the same solution. You can, for example, multiply both sides of an equation by a constant. The equation $x + y = 5$ and the equation $2x + 2y = 10$ are equivalent.

In the same way, there are ways to transform the appearance of a matrix while maintaining the relationships of the numbers within it. These are commonly called *row operations*. You can transform a matrix with the following row operations:

- Multiply or divide a row by a constant.

- Add one row to another, or subtract one row from another.

- Add a multiple of a row to another row.

- Swap the positions of two rows.

It is possible to solve a system of equations by writing an augmented matrix to represent the system, and performing a series of row operations until the equation is in something called *reduced row echelon form*. In this form, each column of the matrix that represents a variable contains a single 1, while the remainder of the column is filled with zeros; the ones are along the diagonal of the matrix. The final, extra column of the augmented matrix is where the solution appears. The following shows you an example of an augmented matrix transformed to reduced row echelon form.

$$\text{Augmented matrix: } \begin{bmatrix} 5 & -3 & | & 11 \\ 2 & 7 & | & 29 \end{bmatrix} \qquad\qquad \text{Reduced row echelon form: } \begin{bmatrix} 1 & 0 & | & 4 \\ 0 & 1 & | & 3 \end{bmatrix}$$

DEFINITION

Reduced row echelon form of a matrix has ones along the diagonal of the row-by-row square portion of the matrix, zeros elsewhere in those columns, and a final column containing the solution.

The original matrix represented the system $5x - 3y = 11$ and $2x + 7y = 29$. If you translate the reduced row echelon form back to a system, it turns out to be $x = 4$ and $y = 3$. The question, of course, is how do you get from augmented form to reduced row echelon form? Having a clear plan is important, so you don't find yourself going in circles.

To put an augmented matrix in reduced row echelon form, follow these steps:

1. Multiply or divide row 1 by a constant to make the coefficient in row 1, column 1 equal to 1.

$$\begin{bmatrix} 5 & -3 & | & 11 \\ 2 & 7 & | & 29 \end{bmatrix} \rightarrow \begin{bmatrix} \frac{5}{5} & \frac{-3}{5} & | & \frac{11}{5} \\ 2 & 7 & | & 29 \end{bmatrix} \rightarrow \begin{bmatrix} 1 & -0.6 & | & 2.2 \\ 2 & 7 & | & 29 \end{bmatrix}$$

2. For each of the other rows in turn, multiply row 1 by the opposite of the number in the target row, column 1. Add that multiple of row 1 to the target row, so the entry in the target row, column 1, becomes 0. Repeat for each row until all other elements in the first column are 0.

$$\begin{bmatrix} 1 & -0.6 & | & 2.2 \\ 2 & 7 & | & 29 \end{bmatrix} \rightarrow \begin{bmatrix} 1 & -0.6 & | & 2.2 \\ 2+1(-2) & 7+(-0.6)(-2) & | & 29+(2.2)(-2) \end{bmatrix} \rightarrow \begin{bmatrix} 1 & -0.6 & | & 2.2 \\ 2-2 & 7+1.2 & | & 29-4.4 \end{bmatrix} \rightarrow \begin{bmatrix} 1 & -0.6 & | & 2.2 \\ 0 & 8.2 & | & 24.6 \end{bmatrix}$$

3. Multiply or divide row 2 by a constant to make the coefficient in row 2, column 2 equal to 1.

$$\begin{bmatrix} 1 & -0.6 & | & 2.2 \\ 0 & 8.2 & | & 24.6 \end{bmatrix} \rightarrow \begin{bmatrix} 1 & -0.6 & | & 2.2 \\ \frac{0}{8.2} & \frac{8.2}{8.2} & | & \frac{24.6}{8.2} \end{bmatrix} \rightarrow \begin{bmatrix} 1 & -0.6 & | & 2.2 \\ 0 & 1 & | & 3 \end{bmatrix}$$

4. Repeat step 2, using multiples of row 2 to change the rest of column 2 to zeros.

$$\begin{bmatrix} 1 & -0.6 & | & 2.2 \\ 0 & 1 & | & 3 \end{bmatrix} \rightarrow \begin{bmatrix} 1+0(0.6) & -0.6+1(0.6) & | & 2.2+3(0.6) \\ 0 & 1 & | & 3 \end{bmatrix} \rightarrow \begin{bmatrix} 1+0 & -0.6+0.6 & | & 2.2+1.8 \\ 0 & 1 & | & 3 \end{bmatrix} \rightarrow \begin{bmatrix} 1 & 0 & | & 4 \\ 0 & 1 & | & 3 \end{bmatrix}$$

5. Repeat, if needed, for column 3 and beyond, until all the columns that represent variables contain zeros and a single 1.

6. The final column contains the values of each variable in the solution.

$$\begin{bmatrix} 1 & 0 & | & 4 \\ 0 & 1 & | & 3 \end{bmatrix} \rightarrow \begin{array}{c} 1x+0y=4 \\ 0x+1y=3 \end{array} \rightarrow \begin{array}{c} x=4 \\ y=3 \end{array}$$

 CALCULATOR CORNER

A graphing calculator that supports matrix operations will likely have row operations commands, but may also have a single command to transform a matrix to reduced row echelon form.

 CHECKPOINT

Find the solution of the system by putting the augmented matrix in reduced row echelon form.

16. $2x - y = 7$
    $x + 3y = -14$

17. $3x + 4y = -2$
    $2x - 2y = 22$

18. $3x - 2y = 4$
    $2x + y = 12$

19. $3x + 2y + 5z = -21$
    $2x + 5y + 4z = 3$
    $-3x - 4y + 2z = -3$

20. $-3x + 4y \quad = 38$
    $5x \quad -3z = -42$
    $3y + 2z = 23$

Solving Systems by Inverse Matrices

When you are asked to divide by a fraction, you actually multiply by its reciprocal. Two numbers are reciprocals if their product is 1, but you probably just think of the reciprocal as the fraction flipped upside down. For example, the reciprocal of $\frac{3}{4}$ is $\frac{4}{3}$ because $\frac{3}{4} \cdot \frac{4}{3} = 1$.

You may have noticed that there was no mention of division in matrix arithmetic. The matrix equivalent of division is multiplying by another matrix, called an *inverse matrix*. In matrix multiplication, the equivalent of a 1 is a square matrix with ones on the major diagonal and zeros elsewhere, called an *identity matrix*. It can be any size necessary for your multiplication, but it must be square. Usually, the identity matrix is denoted [*I*], sometimes with a subscript to indicate its size.

$$I_2 = \begin{bmatrix} 1 & 0 \\ 0 & 1 \end{bmatrix}$$

$$I_3 = \begin{bmatrix} 1 & 0 & 0 \\ 0 & 1 & 0 \\ 0 & 0 & 1 \end{bmatrix}$$

 **DEFINITION**

The **multiplicative inverse of a square matrix** [A] is another square matrix [B] such that [A] · [B] = [B] · [A] = an identity matrix. An **identity matrix** is a square matrix with ones on the major diagonals and zeros elsewhere. Multiplying a matrix and an identity matrix of appropriate size leaves the matrix unchanged.

A matrix [B] is the *multiplicative inverse* of matrix [A] if [A] · [B] = [B] · [A] = [I]. In other words, matrix A and matrix B are inverses if they multiply to an identity matrix, just as two fractions are reciprocals if they multiply to 1. The following shows an example of inverse matrices.

$$\begin{bmatrix} 3 & 1 \\ 5 & 2 \end{bmatrix} \cdot \begin{bmatrix} 2 & -1 \\ -5 & 3 \end{bmatrix} = \begin{bmatrix} 3 \cdot 2 + 1(-5) & 3(-1) + 1 \cdot 3 \\ 5 \cdot 2 + 2(-5) & 5(-1) + 2 \cdot 3 \end{bmatrix} = \begin{bmatrix} 6-5 & -3+3 \\ 10-10 & -5+6 \end{bmatrix} = \begin{bmatrix} 1 & 0 \\ 0 & 1 \end{bmatrix}$$

$$\begin{bmatrix} 2 & -1 \\ -5 & 3 \end{bmatrix} \cdot \begin{bmatrix} 3 & 1 \\ 5 & 2 \end{bmatrix} = \begin{bmatrix} 2 \cdot 3 + (-1)(5) & 2(1) + (-1) \cdot 2 \\ -5 \cdot 3 + 3(5) & -5(1) + 3 \cdot 2 \end{bmatrix} = \begin{bmatrix} 6-5 & 2-2 \\ -15+15 & -5+6 \end{bmatrix} = \begin{bmatrix} 1 & 0 \\ 0 & 1 \end{bmatrix}$$

Finding an Inverse Matrix

Not every matrix has an inverse, so that limits the possibilities for division a bit. In order to have an inverse, a matrix must 1) be square and 2) have a nonzero determinant. The first condition comes from the definition of inverse matrices. In order to be able to multiply in both orders, [A] · [B] and [B] · [A], both matrices must be square. The reason for the second condition will become clear when you go through the process of finding an inverse. While calculating the inverse of a large matrix is a process best left to calculators or computer software, the inverse of a 2-by-2 matrix can be found without that help.

To find the inverse of a 2-by-2 matrix, follow these steps:

1. Find the determinant of the given matrix. (If the determinant is 0, there is no inverse, and you can stop. You'll see why in step 3.) Let's use the matrix for the system $5x - 3y = 11$ and $2x + 7y = 29$ again for the example.

$$\begin{vmatrix} 5 & -3 \\ 2 & 7 \end{vmatrix} = 5 \cdot 7 - 2(-3) = 35 + 6 = 41$$

2. Create a new 2-by-2 matrix by swapping the positions of the numbers on the major diagonal and changing the signs of the numbers on the minor diagonal.

original: $\begin{bmatrix} 5 & -3 \\ 2 & 7 \end{bmatrix}$ new: $\begin{bmatrix} 7 & 3 \\ -2 & 5 \end{bmatrix}$

3. Multiply the matrix you just created by the reciprocal of the determinant you found in step 1 to get the inverse matrix. (Zero has no reciprocal. This is why a matrix with a 0 determinant has no inverse.)

$$\frac{1}{41}\begin{bmatrix} 7 & 3 \\ -2 & 5 \end{bmatrix} = \begin{bmatrix} \frac{7}{41} & \frac{3}{41} \\ \frac{-2}{41} & \frac{5}{41} \end{bmatrix}$$

CALCULATOR CORNER

You can generate an inverse matrix on a calculator once the original matrix is entered, by typing the matrix name followed by the x^{-1} key. Inverses often involve "messy" numbers. Using the calculator's function for converting decimals to fractions may make the inverse more manageable. Your calculator may also have a command designed to create an identity matrix of a specified size automatically. That command only generates the identity matrix. Store the result, or you'll have to regenerate it each time you need it.

CHECKPOINT

Find the inverse of each matrix, if possible.

21. $\begin{bmatrix} 3 & -1 \\ -5 & 2 \end{bmatrix}$

22. $\begin{bmatrix} -1 & 4 & 2 \\ -2 & 2 & -3 \end{bmatrix}$

23. $\begin{bmatrix} 3 & -2 \\ 4 & 1 \end{bmatrix}$

24. $\begin{bmatrix} 1 & 2 & 3 \\ -4 & 5 & 0 \\ 2 & 9 & 3 \end{bmatrix}$

25. $\begin{bmatrix} -3 & 4 & -2 \\ 0 & -1 & 5 \\ -2 & 0 & 3 \end{bmatrix}$

Solving a System by Inverses

Finding an inverse matrix (and dealing with the fractions that often populate inverse matrices) can be worth the effort because it gives you an efficient way to solve a system.

The system of equations can be represented as a matrix equation: the coefficient matrix times a column matrix containing the variables equals the column matrix containing the constants. So, for instance, the matrix equation $\begin{bmatrix} 5 & -3 \\ 2 & 7 \end{bmatrix} \cdot \begin{bmatrix} x \\ y \end{bmatrix} = \begin{bmatrix} 11 \\ 29 \end{bmatrix}$ is equivalent to $\begin{bmatrix} 5x - 3y \\ 2x + 7y \end{bmatrix} = \begin{bmatrix} 11 \\ 29 \end{bmatrix}$ and to the system $5x - 3y = 11$ and $2x + 7y = 29$.

If the matrix equation $\begin{bmatrix} 5 & -3 \\ 2 & 7 \end{bmatrix} \cdot \begin{bmatrix} x \\ y \end{bmatrix} = \begin{bmatrix} 11 \\ 29 \end{bmatrix}$ is multiplied on both sides by the inverse of $\begin{bmatrix} 5 & -3 \\ 2 & 7 \end{bmatrix}$,
you quickly arrive at the solution of the system.

$$\begin{bmatrix} 5 & -3 \\ 2 & 7 \end{bmatrix} \cdot \begin{bmatrix} x \\ y \end{bmatrix} = \begin{bmatrix} 11 \\ 29 \end{bmatrix}$$

$$\begin{bmatrix} \frac{7}{41} & \frac{3}{41} \\ \frac{-2}{41} & \frac{5}{41} \end{bmatrix} \cdot \begin{bmatrix} 5 & -3 \\ 2 & 7 \end{bmatrix} \cdot \begin{bmatrix} x \\ y \end{bmatrix} = \begin{bmatrix} \frac{7}{41} & \frac{3}{41} \\ \frac{-2}{41} & \frac{5}{41} \end{bmatrix} \begin{bmatrix} 11 \\ 29 \end{bmatrix}$$

$$\begin{bmatrix} \frac{35}{41}+\frac{6}{41} & \frac{-21}{41}+\frac{21}{41} \\ \frac{-10}{41}+\frac{10}{41} & \frac{6}{41}+\frac{35}{41} \end{bmatrix} \cdot \begin{bmatrix} x \\ y \end{bmatrix} = \begin{bmatrix} \frac{77}{41}+\frac{87}{41} \\ \frac{-22}{41}+\frac{145}{41} \end{bmatrix}$$

$$\begin{bmatrix} 1 & 0 \\ 0 & 1 \end{bmatrix} \cdot \begin{bmatrix} x \\ y \end{bmatrix} = \begin{bmatrix} 4 \\ 3 \end{bmatrix}$$

$$\begin{bmatrix} x \\ y \end{bmatrix} = \begin{bmatrix} 4 \\ 3 \end{bmatrix}$$

 CALCULATOR CORNER

Enter the matrix of coefficients in a square matrix and the constants in a column matrix. You can find the solution of the system by typing $[A]^{-1}*B$ or whatever the appropriate matrix names may be.

CHECKPOINT

Solve each system by using an inverse matrix.

26. $2x - 3y = 11$
    $x - 4y = 3$

27. $3x - 4y = -4$
    $x + 2y = 12$

28. $3x - 2y = 5$
    $5x + 6y = 13$

29. $x + 2y + 3z = 14$
    $x - 2y + 2z = 3$
    $x + 2y - 4z = -7$

30. $x - y + z = 15$
    $x + 2y + 3z = 100$
    $x - 4y + 5z = 50$

The Least You Need to Know

- Matrices are rectangular arrangements of numbers that can be added, subtracted, and multiplied. It is also possible to find an inverse of a square matrix.

- A system of equations can be represented by a square matrix of coefficients and a column matrix of constants or by a rectangular matrix formed by annexing the column of constants to the square matrix.

- Cramer's rule creates a square matrix of coefficients and modifies copies of that matrix for each variable, replacing the column containing coefficients of that variable with the constants. The value of each variable is equal to the determinant of its modified matrix divided by the determinant of the matrix of coefficients.

- Systems can also be solved by multiplying the inverse of the matrix of coefficients times the matrix of constants, or by using row operations to reduce the augmented matrix to reduced row echelon form.

Quadratic Relationships

Your first experiences with quadratic equations were limited by the tools you had available to you. You might have taken a square root now and then, but for most people, solving quadratics meant factoring or the quadratic formula.

Did you wonder about the quadratic formula when you first saw it? Where did it come from, and why did it work? This part of the journey is the time to explore the idea of roots and radicals more fully, to trace the development of the different methods of solving quadratics, and to imagine what might happen if y were squared instead of or in addition to x.

Roots and Radicals

In a first course in algebra, you learn that square roots are the opposite, or inverse, of the second power. Although you connect roots to powers in that way, radicals seem to have their own system of arithmetic, distinct from the rules you learned for working with exponents. Algebra II is the time to make a few important shifts. In this chapter, you learn to move radicals into the system of exponents, extend the idea of powers and roots to higher powers, and view roots as a class of functions in their own right.

In This Chapter

- Expressing roots with exponents
- Solving equations that involve radicals
- Graphing radical equations by transformations

Rational Exponents

Algebra I tells you that $\left(\sqrt{x}\right)^2 = x$ and $\sqrt{x^2} = |x|$. (The absolute value sign pops up in the second because x^2 will be positive, and its square root will be positive, even if the original value of x was negative.) If there were an exponent that meant "the square root of," what would it be? Let's call it r for now and see if the rules of exponents will give a clue.

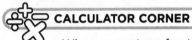

CALCULATOR CORNER

When you enter a fractional exponent on a calculator, enclose the exponent in parentheses. If you type 8^2/3, you'll get $\frac{64}{3}$. Typing 8^(2/3) will get you the correct answer of $8^{\frac{2}{3}} = 2^2 = 4$.

If x^r means $\sqrt{x}$, then $(x^r)^2 = x$. According to the rules for exponents, you multiply those exponents, so $x^{2r} = x$. That means $2r$ must equal 1, so r must be $\frac{1}{2}$. Will $\left(x^2\right)^{\frac{1}{2}} = |x|$? You'll have to understand why the absolute value is necessary, but the rest of this works, so it seems that a square root can be expressed as a $\frac{1}{2}$ exponent.

That discovery allows you to extend the idea of a fractional exponent to other roots. The cube root can be expressed as the $\frac{1}{3}$ power because $\left(x^3\right)^{\frac{1}{3}} = 1$. In general, for any power n, $\left(x^n\right)^{\frac{1}{n}} = x$, so the nth root can be represented by the $\frac{1}{n}$ power.

Being able to write roots as exponents allows you to use a single set of rules to simplify expressions involving both powers and roots. A fraction used as an exponent is called a *rational exponent*. The numerator represents a power and the denominator a root. You can simplify a rational exponent as you would simplify a fraction, and often this makes your calculation simpler without changing the result. If you need to raise $\sqrt[3]{2}$ to the sixth power, for example, $\left(\sqrt[3]{2}\right)^6 = \left(2^{\frac{1}{3}}\right)^6 = 2^{\frac{6}{3}} = 2^2 = 4$. Or if you want to find the square root of 2^{11}, $\sqrt{2^{11}} = 2^{\frac{11}{2}} = 2^{\frac{10}{2}} \cdot 2^{\frac{1}{2}} = 2^5 \cdot 2^{\frac{1}{2}} = 32\sqrt{2}$.

Like any other calculation involving roots, you must be aware that in the real numbers, you can only take an even root—square root, fourth root, and so on—of a non-negative number. But once you've verified that you're working on an acceptable domain, you can choose to do the power and root in any convenient order. It's generally simpler to do roots first, because the numbers get smaller, before raising to a power makes them bigger. So $16^{\frac{3}{4}}$ is easier if you take the fourth root of 16 to get 2 and then raise that to the third power, yielding 8, than to raise 16 to the third and then try to find the fourth root.

CHECKPOINT

Evaluate each expression.

1. $8^{\frac{5}{3}}$

2. $(-32)^{\frac{2}{5}}$

3. $\left(\frac{16}{25}\right)^{\frac{1}{2}}$

4. $10,000^{\frac{3}{4}}$

5. $(-343)^{\frac{4}{3}}$

Simplifying Radical Expressions

Simplifying radical expressions in general, whether written as radicals or using *rational exponents*, follows the same rules as for square roots. You want the smallest possible number under the radical, and you don't want to leave a radical in the denominator.

CALCULATOR CORNER

If your calculator has a key like $\sqrt[n]{x}$, you usually enter the value of n, the index of the radical, and then hit the key, followed by the value of x. However, you can easily enter the same thing as $x^\wedge(1/n)$.

To simplify $\sqrt{175x^3y^5}$, for example, you look for factors of the radicand that are perfect squares.

$$\sqrt{175x^3y^5} = \sqrt{25 \cdot 7 \cdot x^2 \cdot x \cdot y^4 \cdot y} = 5|x|y^2\sqrt{7xy}$$

In similar fashion, you simplify $\sqrt[3]{40a^4b^7}$ by looking for factors that are perfect cubes.

$$\sqrt[3]{40a^4b^7} = \sqrt[3]{8 \cdot 5 \cdot a^3 \cdot a \cdot b^6 \cdot b} = 2ab^2\sqrt[3]{5ab}$$

Another part of simplifying a radical expression may involve *rationalizing the denominator*. This process is pretty simple for single terms. For instance, to rationalize a denominator with a single square root, multiply the numerator and denominator by that square root.

$$\frac{3}{\sqrt{5}} \cdot \frac{\sqrt{5}}{\sqrt{5}} = \frac{3\sqrt{5}}{5}$$

To rationalize a denominator with a single cube root, you need to multiply the numerator and denominator by the square of that cube root, because you need to end up with a cube root to the third power.

$$\frac{8}{\sqrt[3]{4}} \cdot \frac{\left(\sqrt[3]{4}\right)^2}{\left(\sqrt[3]{4}\right)^2} = \frac{8\left(\sqrt[3]{4}\right)^2}{\left(\sqrt[3]{4}\right)^3} = \frac{8\left(\sqrt[3]{4}\right)^2}{4} = 2\left(\sqrt[3]{4}\right)^2$$

Before you leave that example, let's try to simplify that remaining radical. Rational exponents may help.

$$2\left(\sqrt[3]{4}\right)^2 = 2 \cdot 4^{\frac{2}{3}} = 2 \cdot 16^{\frac{1}{3}} = 2 \cdot (8 \cdot 2)^{\frac{1}{3}} = 2 \cdot 2 \cdot 2^{\frac{1}{3}} = 4\sqrt[3]{2}$$

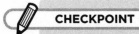

CHECKPOINT

Simplify each expression.

6. $\sqrt[3]{8x^4y^5}$

7. $\sqrt{27a^5b^4c^7}$

8. $\left(\frac{81x^7y^5}{16z^9}\right)^{\frac{1}{3}}$

9. $\sqrt[5]{\frac{64x^7y^8}{z^{10}}}$

10. $\left(\frac{27x^3y^4}{8z^3}\right)^{\frac{5}{2}}$

Rationalizing More Complex Denominators

Rationalizing a denominator with more than one term requires a slightly different strategy. If the denominator is a sum or difference involving one or more square roots, you need to multiply the numerator and denominator by the *conjugate* of the denominator. If the denominator is a sum, the conjugate is the difference of the same two terms. For example, the conjugate of $8 + \sqrt{3}$ is $8 - \sqrt{3}$.

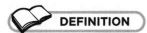

DEFINITION

A **rational exponent** is a fraction, $\frac{p}{r}$, in which the numerator, p, indicates a power to which the base should be raised, and the denominator, r, denotes a root to be taken. **Rationalizing the denominator** is the process of transforming an algebraic fraction so that no radicals remain in the denominator. If the denominator is a sum of two terms, one or both of which is a radical, rationalizing requires multiplying by the **conjugate,** or the difference of the same two terms. The conjugate of a difference of two terms involving a radical is the sum of those two terms.

When you multiply a binomial by its conjugate, the result is a difference of squares: $\left(8 + \sqrt{3}\right)\left(8 - \sqrt{3}\right) = 8^2 - \sqrt{3}^2 = 64 - 3 = 61$. So when you multiply the numerator and denominator by the conjugate of the denominator, the denominator is rationalized.

$$\frac{6}{11+\sqrt{5}} \cdot \frac{11-\sqrt{5}}{11-\sqrt{5}} = \frac{6\left(11-\sqrt{5}\right)}{11-5} = \frac{6\left(11-\sqrt{5}\right)}{6} = 11 - \sqrt{5}$$

Rationalizing a denominator that is a sum or difference involving a cube root is a bit more complicated, because just multiplying by the conjugate won't eliminate the cube root. You need to rely instead on factoring patterns commonly seen when working with cubic expressions.

Sum of cubes: $a^3 + b^3 = (a + b)(a^2 - ab + b^2)$

Difference of cubes: $a^3 - b^3 = (a - b)(a^2 + ab + b^2)$

So if you have a cube root in the denominator, you want to get to either a sum of cubes or a difference of cubes. To rationalize $\frac{4}{3+\sqrt[3]{2}}$, for instance, you need to see $3+\sqrt[3]{2}$ as the $a + b$ factor in the sum of cubes rule. Multiply the numerator and denominator by the equivalent $a^2 - ab + b^2$—in this case, $3^2 - 3\sqrt[3]{2} + \sqrt[3]{2}^2$. Take your time. It will simplify.

$$\frac{4}{3+\sqrt[3]{2}} \cdot \frac{3^2-3\sqrt[3]{2}+\sqrt[3]{2}^2}{3^2-3\sqrt[3]{2}+\sqrt[3]{2}^2} = \frac{4\left(3^2-3\sqrt[3]{2}+\sqrt[3]{2}^2\right)}{3^3+\sqrt[3]{2}^3} = \frac{4\left(9-3\sqrt[3]{2}+\sqrt[3]{4}\right)}{27+2} = \frac{4\left(9-3\sqrt[3]{2}+\sqrt[3]{4}\right)}{29}$$

The reason you don't see many of these problems, of course, is that even though you've rationalized the denominator, the numerator is still unwieldy. As calculators become more common, rationalizing denominators like these becomes less crucial, but it's good to have that ability in your skill set.

CHECKPOINT

Rationalize each denominator and leave the expression in simplest form.

11. $\frac{5}{2+\sqrt{3}}$

12. $\frac{3}{\sqrt{2}-9}$

13. $\frac{2+3\sqrt{5}}{8-\sqrt{5}}$

14. $\frac{2}{\sqrt[3]{5}}$

15. $\frac{2+\sqrt[3]{6}}{\sqrt[3]{3}}$

Solving Radical Equations

Although the use of rational exponents makes numeric calculations involving roots easier, algebra is most concerned with situations in which variables appear under the radical or are raised to a fractional power. The basic rule still applies for solving radical equations—isolate the radical and then raise both sides of the equation to the appropriate power to eliminate the root. Take note that if you have cube roots, you must raise both sides to the third power. Raising both sides of an equation to a power larger than 3 can quickly become tedious, so you might turn to other methods—like solving graphically on a calculator—for those. Check your solutions in the original equations because *extraneous solutions* are possible.

DEFINITION

An **extraneous solution** of an equation is a solution, which is produced by legitimate algebraic techniques, that does not satisfy the original equation. This can occur when a transformation, such as squaring both sides, allows the inclusion of values not in the original domain.

Let's try an example, $\sqrt[3]{7+4x}-5=-3$. To start, isolate the radical by adding 5 to both sides.

$$\sqrt[3]{7+4x}\underset{+5}{-5}=\underset{+5}{-3}$$
$$\sqrt[3]{7+4x}=2$$

Raise both sides to the third power.

$$\sqrt[3]{7+4x}^3=2^3$$
$$7+4x=8$$

Solve the resulting equation.

$$7+4x=8$$
$$7+4x-7=8-7$$
$$4x=1$$
$$x=\tfrac{1}{4}$$

Checking, you see that $\sqrt[3]{7+4\left(\tfrac{1}{4}\right)}=\sqrt[3]{7+1}=2$, so the solution is accepted.

When more than one radical is present, isolate one radical, raise both sides to the appropriate power, and simplify. You then isolate the remaining radical and repeat the process. For instance, to solve $\sqrt{2x-7}+\sqrt{6x+1}=6$, begin by isolating one radical.

$$\sqrt{2x-7}=6-\sqrt{6x+1}$$

Square both sides, remembering that the right side must be FOILed (see Chapter 1 for a refresher on FOIL).

$$\sqrt{2x-7}^2=\left(6-\sqrt{6x+1}\right)^2$$
$$2x-7=36-12\sqrt{6x+1}+6x+1$$

Simplify and isolate the remaining radical.

$$2x-7=36-12\sqrt{6x+1}+6x+1$$
$$2x-7=6x+37-12\sqrt{6x+1}$$
$$12\sqrt{6x+1}=4x+44$$
$$\sqrt{6x+1}=\tfrac{4x+44}{12}=\tfrac{x+11}{3}$$

Square both sides again.

$$\sqrt{6x+1}^{\,2} = \left(\tfrac{x+11}{3}\right)^2$$
$$6x+1 = \tfrac{x^2+22x+121}{9}$$

Solve the equation. This one will be a quadratic equation.

$$6x+1 = \tfrac{x^2+22x+121}{9}$$
$$54x+9 = x^2+22x+121$$
$$x^2-32x+112 = 0$$
$$(x-4)(x-28) = 0$$
$$x-4 = 0 \qquad x-28 = 0$$
$$x = 4 \qquad\quad x = 28$$

If $x = 4$, $\sqrt{2\cdot 4-7}+\sqrt{6\cdot 4+1} = \sqrt{1}+\sqrt{25} = 6$, so that solution is acceptable. If $x = 28$, $\sqrt{2\cdot 28-7}+\sqrt{6\cdot 28+1} = \sqrt{56-7}+\sqrt{168+1} = \sqrt{49}+\sqrt{169} = 7+13 \neq 6$, so reject this solution as extraneous.

Let's look at one more example, this time with a fourth root: $\sqrt[4]{5x^2-6} = x$. The radical is already isolated, so raise both sides to the fourth power.

$$\sqrt[4]{5x^2-6}^{\,4} = x^4$$
$$5x^2-6 = x^4$$
$$x^4-5x^2+6 = 0$$

Although this is a fourth-degree equation, it has the form of a quadratic, which is clearer if you substitute a for x^2 and a^2 for x^4. The equation becomes $a^2 - 5a + 6 = 0$ and can be solved by factoring.

$$a^2-5a+6 = 0$$
$$(a-3)(a-2) = 0$$
$$a-3 = 0 \qquad a-2 = 0$$
$$a = 3 \qquad\quad a = 2$$

 CALCULATOR CORNER

Graphing two functions—one for the left side and one for the right side of your equation—on a graphing calculator and counting points of intersection gives you a quick way to know if you have extraneous solutions.

Remember that a took the place of x^2, and these solutions become $x^2 = 3$ and $x^2 = 2$, giving four solutions: $x = \pm\sqrt{3}$ and $x = \pm\sqrt{2}$. A check will reveal that the negative solutions must be rejected. The fourth root will produce only positive values.

CHECKPOINT

Solve each equation.

16. $\sqrt{8x-4} + 2\sqrt{x-1} = 10$

17. $\sqrt[3]{4+5x} + 7 = 11$

18. $\sqrt{2x-5} = 1 - \sqrt{3x-5}$

19. $\sqrt[3]{48-3x} + 3 = 6$

20. $\sqrt[4]{8x^2 - 15} = x$

Graphing Radical Functions

The parent function for the square root family, shown in the following graph, is defined for values of $x \geq 0$ and produces values of $y \geq 0$. The restriction on the domain exists because, in the real numbers, there is no square root of a negative number. The range includes only non-negative values of y because the standard agreement is that the square root sign denotes the principal or positive square root. If that agreement were not in place, there would be no square root function, because each x (other than 0) would have both a positive and a negative square root.

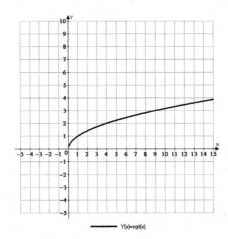

Y(x)=sqrt(x)

The shape of the graph is that of one side of a parabola. To transform the graph using key points, you could use (0, 0), (1, 1), and (4, 2). All the rules for transformations remain as you've seen them in previous chapters.

To graph $f(x) = \sqrt{x-3} + 5$, you would shift the parent graph 3 units right and 5 units up.

Key Points	3 Right	5 Up
(0, 0)	(3, 0)	(3, 5)
(1, 1)	(4, 1)	(4, 6)
(4, 2)	(7, 2)	(7, 7)

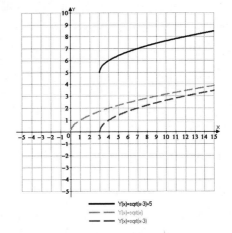

The function $g(x) = 2\sqrt{-x+4}$ shifts the parent 4 units left, reflects the graph across the y-axis, and stretches it vertically by a factor of 2.

Key Points	4 Left	Reflect	Vertical Stretch
(0, 0)	(−4, 0)	(4, 0)	(4, 0)
(1, 1)	(−3, 1)	(3, 1)	(3, 2)
(4, 2)	(0, 2)	(0, 2)	(0, 4)

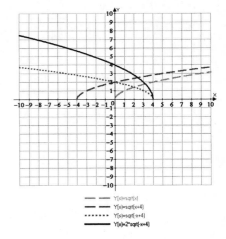

The parent function for the cube root family is defined for all real numbers, because you can find the cube root of a negative number, and its range includes all real numbers, even though the graph climbs and declines very slowly. If you follow it to larger and larger (or smaller and smaller) values of x, it continues to increase (or decrease).

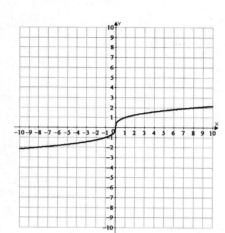

The key points for the cube root graph could be (0, 0), (1, 1), and (−1, −1), but if you want to move farther from the origin, (8, 2) and (−8, −2) could also be used.

The function $f(x) = -\frac{1}{2}(x+2)^{\frac{1}{3}} - 1$ is a member of the cube root family; the $\frac{1}{3}$ exponent replaces the $\sqrt[3]{\ }$. This graph shifts the parent 2 units left, compresses it vertically, reflects it across the x-axis, and shifts it 1 unit down.

Key Points	2 Left	Compress	Reflect	1 Down
(0, 0)	(−2, 0)	(−2, 0)	(−2, 0)	(−2, −1)
(8, 2)	(6, 2)	(6, 1)	(6, −1)	(6, −2)
(−8, −2)	(−10, −2)	(−10, −1)	(−10, 1)	(−10, 0)

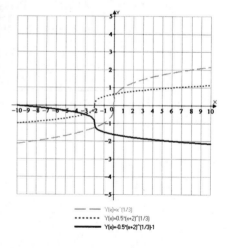

- - - - Y(x)=x^(1/3)
········ Y(x)=0.5*(x+2)^(1/3)
——— Y(x)=-0.5*(x+2)^(1/3)-1

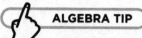

ALGEBRA TIP

Graphs of other even root functions resemble the square root function, and other odd roots look like the cube root. The difference is in the curvature of the graph. A fourth root increases more slowly than a square root, for example, but the basic shape is similar.

CHECKPOINT

Sketch a graph of each function.

21. $f(x) = \sqrt{x-5} + 3$

22. $y = 2 - 3\sqrt{x+4}$

23. $f(x) = \frac{1}{2}\sqrt{4-x} - 5$

24. $g(x) = \sqrt[3]{x+4} - 5$

25. $y = 5 - 2\sqrt[3]{x+4}$

The Least You Need to Know

- Rational exponents indicate both a power and a root. The numerator is the power and the denominator the root.

- Simplify radicals by finding factors that are perfect squares for square roots, or perfect cubes for third roots. Take those roots, leaving the smallest expression possible under the radical.

- Rationalize a denominator that is a single term involving a square root by multiplying the numerator and denominator by that radical. Rationalize a denominator that is a sum or difference involving a square root by multiplying the numerator and denominator by the conjugate of the denominator.

- Solve a radical equation by isolating the radical and raising both sides to a power that will cancel the root. If there is more than one radical, isolate one, raise both sides to a power, isolate the remaining radical, and raise to a power again.

- Always check for extraneous solutions. Extraneous solutions can occur when a transformation, such as squaring both sides, allows the inclusion of values not in the original domain.

- Graph square root or cube root functions by transforming the parent functions.

Quadratics

Most of your work in algebra involves solving equations, and most of those equations are either linear or quadratic. Algebra II introduces a method of solving quadratic equations that you probably didn't see in algebra I—a method that explains the quadratic formula you probably did learn. This chapter teaches you this solving method, removes one of the principal obstacles you've met (the inability to find the square root of a negative number), and opens up a new system of numbers, called the *complex numbers*.

Quadratic Equations

A *quadratic equation* is an equation of the form $ax^2 + bx + c = 0$, where a, b, and c are real numbers and a is not 0. For example, the equation $3x^2 - 2x + 5 = 0$ has $a = 3$, $b = -2$, and $c = 5$. Allowing a to be 0 would make the x^2-term disappear, eliminating the principal characteristic of a quadratic equation. You can still have a quadratic equation with the other terms missing, but not without a squared term.

In This Chapter

- Solving quadratic equations by completing the square
- Using the quadratic formula
- Working with complex numbers
- Graphing quadratic functions
- Fitting quadratic equations to data

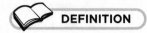 **DEFINITION**

A **quadratic equation** is an equation of the form $ax^2 + bx + c = 0$, in which a, b, and c are real numbers and a is not 0.

Square Root Method

When the x-term disappears, because $b = 0$, the equation has the form $ax^2 + c = 0$ and can be solved by isolating the x^2 and taking the square root of both sides. As you do this, remember to consider both the positive and negative square root.

To solve $4x^2 - 25 = 0$, add 25 to both sides, divide by 4, and take the square root.

$$4x^2 - 25 = 0$$
$$4x^2 = 25$$
$$x^2 = \frac{25}{4}$$
$$x = \pm\sqrt{\frac{25}{4}} = \pm\frac{5}{2}$$

The square root method can also be used to solve a quadratic equation in which b is not 0, if the quadratic expression can be rewritten as the square of a binomial. For example, the equation $4x^2 + 4x + 1 = 9$ can be written as $(2x + 1)^2 = 9$. Once you do that, you can solve it by taking the square root of both sides, as shown in the following.

$$(2x+1)^2 = 9$$
$$2x+1 = \pm 3$$

$$2x+1 = 3 \qquad 2x+1 = -3$$
$$2x = 2 \qquad 2x = -4$$
$$x = 1 \qquad x = -2$$

As you can imagine, it's not every quadratic equation that can be rewritten as the square of a binomial, but there is a way to transform the equation to achieve that.

 CHECKPOINT

Solve by the square root method.

1. $5x^2 - 80 = 0$ 3. $(3x - 7)^2 = 25$ 5. $(9 + 3y)^2 - 11 = 214$

2. $\frac{1}{2}x^2 - \frac{25}{8} = 0$ 4. $(4 - 5x)^2 = 48$

Completing the Square

Completing the square is a method for transforming a quadratic equation so that one side is a perfect square trinomial and the other side is a constant. It allows you to solve any quadratic equation by the square root method (though it can lead to some unpleasant fractions).

DEFINITION

> **Completing the square** is a method of solving quadratic equations by adjusting the constant term of the quadratic polynomial so it becomes a perfect square trinomial. It can then be written as the square of a binomial and solved by the square root method.

The key, of course, is that you need to know what a perfect square trinomial looks like. Technically, that's $a^2x^2 + 2abx + b^2$, the pattern you produce when you square $ax + b$. For instance, squaring $3x - 5$ will give you $9x^2 - 30x + 25$. You can get through the process of completing the square, however, if you know the simpler version, which occurs only when $a = 1$: $x^2 + 2bx + b^2$. Square $x + 6$ and you get $x^2 + 12x + 36$. To make completing the square a little easier, anytime the coefficient of the squared term is not 1, you'll divide through the equation by that coefficient before you begin.

Because you want one side of the equation to be a constant, it isn't wise to try to adjust the terms that contain variables. If you add or subtract an x-term to one side, you have to do the same on the other side. To keep one side a constant, you'll focus on adjusting only the constants. Here's the process, step by step, with a simple example to start:

1. Move any constant terms to one side of the equation, keeping the x^2-term and x-term on the other side.

$$4x^2 - 24x + 2 = -30$$
$$4x^2 - 24x = -32$$

2. Divide both sides of the equation by the coefficient of the x^2-term. Dividing through is what often creates fractions, but be patient.

$$4x^2 - 24x = -32$$
$$\frac{4x^2 - 24x}{4} = \frac{-32}{4}$$
$$x^2 - 6x = -8$$

3. This is the moment when you actually make the variable side a perfect square trinomial. You want to match your quadratic expression to $x^2 + 2bx + b^2$. You have the x^2, you have some coefficient in front of x, and you're going to create your own constant term. The coefficient of x is $2b$, so take half of that coefficient, square it, and add it to both sides.

$$x^2 + 2b + b^2$$
$$x^2 - 6x + ? = -8$$
$$2b = -6 \rightarrow b = -3 \rightarrow b^2 = 9$$
$$x^2 - 6x + 9 = -8 + 9$$

4. Rewrite the quadratic expression as $(x + b)^2$.

$$x^2 - 6x + 9 = -8 + 9$$
$$(x - 3)^2 = 1$$

5. Solve using the square root method.

$$(x - 3)^2 = 1$$
$$x - 3 = \pm\sqrt{1}$$
$$x - 3 = 1 \qquad x - 3 = -1$$
$$x = 4 \qquad\quad x = 2$$

Here's another example, one where you'll find both fractions and a constant that is not a perfect square.

Solve by completing the square: $3x^2 + 5x + 9 = 8$.

1. Move the constant.

$$3x^2 + 5x = -1$$

2. Divide through by a.

$$x^2 + \tfrac{5}{3}x = -\tfrac{1}{3}$$

3. Take half the coefficient of x, square it, and add it to both sides.

$$x^2 + \tfrac{5}{3}x + \left(\tfrac{5}{6}\right)^2 = -\tfrac{1}{3} + \left(\tfrac{5}{6}\right)^2$$

4. Rewrite the quadratic expression as $(x + b)^2$.

$$x^2 + \tfrac{5}{3}x + \left(\tfrac{5}{6}\right)^2 = -\tfrac{1}{3} + \left(\tfrac{5}{6}\right)^2$$
$$\left(x + \tfrac{5}{6}\right)^2 = -\tfrac{1}{5} + \tfrac{25}{36} = \tfrac{24}{36}$$

5. Solve by the square root method.

$$x + \tfrac{5}{6} = \pm\sqrt{\tfrac{24}{36}}$$

$$x + \tfrac{5}{6} = \pm\tfrac{\sqrt{24}}{6}$$

$$x = -\tfrac{5}{6} \pm \tfrac{\sqrt{24}}{6}$$

$$x = \tfrac{-5 \pm 2\sqrt{6}}{6}$$

You can leave that expression in simplest radical form for an exact answer or use your calculator for a decimal approximation.

CHECKPOINT

Solve by completing the square.

6. $x^2 - 15x + 2 = 0$

7. $y^2 - 72y + 8 = 0$

8. $2x^2 + 12x - 3 = 0$

9. $3x^2 - 12x + 1 = 0$

10. $5x^2 + 13x + 2 = 0$

Quadratic Formula

If the process of completing the square isn't your idea of great fun, you'll be pleased to know that as early as the twelfth century, mathematicians were looking for a simpler way to solve quadratic equations. What they handed down to us is a formula—the *quadratic formula*—that is the result of solving the general form of the quadratic equation, $ax^2 + bx + c = 0$, by completing the square.

DEFINITION

The **quadratic formula,** $x = \frac{-b \pm \sqrt{b^2 - 4ac}}{2a}$, is a formula that produces the solutions of a quadratic equation, $ax^2 + bx + c = 0$, when the values of a, b, and c are substituted into the formula and simplified.

To derive the quadratic formula, you solve $ax^2 + bx + c = 0$ by completing the square.

1. Move the constant of the quadratic equation.

$$ax^2 + bx = -c$$

2. Divide through by a.

$$x^2 + \tfrac{b}{a}x = -\tfrac{c}{a}$$

3. Take half the coefficient of x, square it, and add it to both sides.

$$x^2 + \frac{b}{a}x + \left(\frac{b}{2a}\right)^2 = -\frac{c}{a} + \left(\frac{b}{2a}\right)^2$$

4. Rewrite the quadratic expression as $(x + b)^2$.

$$\left(x + \frac{b}{2a}\right)^2 = -\frac{c}{a} + \left(\frac{b}{2a}\right)^2$$

$$\left(x + \frac{b}{2a}\right)^2 = -\frac{c}{a} + \frac{b^2}{4a^2}$$

$$\left(x + \frac{b}{2a}\right)^2 = -\frac{4ac}{4a^2} + \frac{b^2}{4a^2}$$

$$\left(x + \frac{b}{2a}\right)^2 = \frac{b^2 - 4ac}{4a^2}$$

5. Solve.

$$x + \frac{b}{2a} = \pm\sqrt{\frac{b^2 - 4ac}{4a^2}}$$

$$x + \frac{b}{2a} = \frac{\pm\sqrt{b^2 - 4ac}}{2a}$$

$$x = \frac{-b \pm \sqrt{b^2 - 4ac}}{2a}$$

The result, $x = \frac{-b \pm \sqrt{b^2 - 4ac}}{2a}$, allows you to substitute known values for a, b, and c and simplify to arrive at the solutions of the equation.

As an example, let's solve $5x^2 + 7x - 3 = 0$. You must have all nonzero terms on one side and arranged in descending order of exponent: squared term, first power, and constant. To start, identify the values of a, b, and c. In this example, $a = 5$, $b = 7$, and $c = -3$.

Write the formula $x = \frac{-b \pm \sqrt{b^2 - 4ac}}{2a}$, and replace a, b, and c with the values specific to this problem.

$$x = \frac{-7 \pm \sqrt{7^2 - 4 \cdot 5 \cdot (-3)}}{2 \cdot 5}$$

Simplify under the radical first, and then simplify the denominator.

$$x = \frac{-7 \pm \sqrt{49 + 60}}{2 \cdot 5} = \frac{-7 \pm \sqrt{109}}{2 \cdot 5} = \frac{-7 \pm \sqrt{109}}{10}$$

CALCULATOR CORNER

Consider writing a calculator program for the quadratic formula. First, input the values of a, b, and c. Store and then display—or just display—two calculations: $\frac{-b + \sqrt{b^2 - 4ac}}{2a}$ and $\frac{-b + \sqrt{b^2 + 4ac}}{2a}$. If you're not ready to write it yourself, you can download a version from various sites on the internet.

In some circumstances, you may be able to simplify the radical, and you may find a common factor in the numerator and denominator that can be cancelled, but for this example, $x = \frac{-7 \pm \sqrt{109}}{10}$ is your simplest answer. You can still use your calculator to find decimal approximations of $x = \frac{-7+\sqrt{109}}{10}$ and $x = \frac{-7-\sqrt{109}}{10}$, if that's helpful.

CHECKPOINT

Solve by the quadratic formula.

11. $2x^2 + 3 = 12x$ 14. $8x^2 - 9x - 2 = 0$

12. $2 = 4x - x^2$ 15. $7x^2 - 7x + 1 = 0$

13. $30z^2 - 17z = 0$

One of the advantages of the quadratic formula as a method for solving quadratic equations is that a part of it, called the *discriminant,* gives you information about the number and type of solutions even before you complete the solution. The radicand—the expression under the radical—is the discriminant in the quadratic formula: $b^2 - 4ac$. The following are what the discriminant can tell you about the number and type of solutions:

- If $b^2 - 4ac > 0$, there will be two real solutions, which will be rational numbers if the discriminant is a perfect square and irrational otherwise.

- If $b^2 - 4ac = 0$, you will see only one solution, $x = \frac{-b}{2a}$. In fact, this is what is commonly called a *double root,* because it is actually the same solution occurring twice. The more formal way to say this is that you have a single solution with a *multiplicity* of 2.

- If $b^2 - 4ac < 0$, there are no real solutions, because in the real numbers, it is impossible to find the square root of a negative number.

DEFINITION

The **discriminant** is $b^2 - 4ac$, the radicand of the quadratic formula. The sign of the discriminant tells you whether the quadratic equation has two real solutions ($b^2 - 4ac > 0$), one real solution ($b^2 - 4ac = 0$), or two nonreal solutions ($b^2 - 4ac < 0$). A **double root** is a solution that occurs twice for the same equation. This is also known as a solution with a multiplicity of 2. The **multiplicity** of a solution is the number of times a solution occurs and corresponds to the power to which the corresponding factor is raised.

And that, in elementary algebra, is the end of the matter. If you started to solve a quadratic equation and found the discriminant was negative, you said there was no solution and you stopped. But for those who can conceive of a system of numbers larger than the real numbers, there's more to say.

CHECKPOINT

Determine the number and type of solutions, but do not solve.

16. $4x^2 + 7x + 1 = 0$ 19. $9x^2 + 12x + 4 = 0$

17. $2x^2 - 3x - 3 = 0$ 20. $25x^2 - 36 = 0$

18. $5x^2 + 2x + 3 = 0$

Complex Numbers and Their Arithmetic

The first step into this wider world is defining the *imaginary unit,* traditionally called *i,* as the square root of –1. From this imaginary unit, we build a system of numbers that are real number multiples of *i* and call this system the *imaginary numbers.* The square root of a negative number can be expressed as an imaginary number by the same process you use to get to simplest radical form, as you can see with these examples.

$$\sqrt{-25} = \sqrt{25(-1)} = \sqrt{25}\sqrt{-1} = 5i \quad \text{and} \quad \sqrt{-48} = \sqrt{16 \cdot 3 \cdot (-1)} = 4i\sqrt{3}$$

When you combine the real numbers with the imaginary numbers, you create a set known as the *complex number.* Each complex number can be written in the form $a + bi$, where a is a real number and bi is an imaginary number. The real numbers are a subset of the complex numbers, because every real number can be written as $a + 0i$. Likewise, the imaginary numbers are also a subset of the complex numbers, because every imaginary number can be written as $0 + bi$.

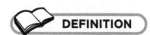

DEFINITION

The **imaginary unit,** *i,* is the symbol for the square root of –1. An **imaginary number** is a multiple of *i.* A **complex number** is a number of the form $a + bi$, where a is a real number and bi is an imaginary number.

Doing arithmetic with complex numbers calls upon many things you learned working with variables. Addition and subtraction are very much about combining like terms, as you can see in the following examples.

Addition: $(2 + 3i) + (5 - 2i) = 7 + i$

Subtraction: $(2 + 3i) - (5 - 2i) = -3 + 5i$

Multiplication can be accomplished with the distributive property or the FOIL rule, as shown in the following examples. To simplify the result properly, remember that $i^2 = \sqrt{-1}^2 = -1$.

Multiplication: $6i(-4 + 5i) = -24i + 30i^2 = -30 - 24i$

$(2 - 3i)(3 + 5i) = 6 + 10i - 9i - 15i^2 = 6 + 15 + i = 21 + i$

For division, remember that i was defined as a square root and call upon your work with radicals. Multiply the numerator and denominator by the conjugate of the denominator; you can see this in the following example.

Division: $\frac{2+4i}{5-3i} \cdot \frac{5+3i}{5+3i} = \frac{10+6i+20i+12i^2}{25+15i-15i-9i^2} = \frac{-2+26i}{34} = -\frac{1}{17} + \frac{13}{17}i$

With the introduction of complex numbers, every quadratic equation has a solution. Now a negative discriminant still tells you there are no real solutions, but there are two complex solutions.

 CALCULATOR CORNER

Make sure your calculator is set to display and calculate with complex numbers. If entering $\sqrt{-1}$ gives you an error, check the mode or settings menu to change from real to $a + bi$ mode.

 CHECKPOINT

Perform the indicated operations and leave the expression in simplest form.

21. $(3 - 2i) + (4 + 9i) - (2 - 5i)$

22. $(2 - 5i)(3 + 4i)$

23. $\frac{6-5i}{3+2i}$

Solve each equation, putting complex solutions in simplest form.

24. $x^2 + 4x + 7 = 0$ 25. $3x^2 + 8x + 12 = 0$

Graphing Quadratic Functions

The *x*-intercepts of any graph correspond to the solutions of its corresponding equation. When graphing a quadratic function, you can transform the parent graph, the parabola, and you can also use the solutions of the equation to set the *x*-intercepts. There will be two *x*-intercepts if the discriminant of the corresponding equation is positive. But if the discriminant is 0, only the vertex, or turning point, of the parabola will touch the *x*-axis. A parabola that floats entirely above or entirely below the *x*-axis corresponds to an equation with nonreal solutions.

Key points for the parent parabola might be the vertex (0, 0) and (1, 1) and (–1, 1), or you might also use (2, 4) and (–2, 4) to help set the curvature of the parabola. Transformations apply to the parabola as to any other graph family, but knowing the vertex and the intercepts may be enough.

The domain of any quadratic function is the entire set of real numbers, but the range is always restricted. It runs from the *y*-coordinate of the vertex up, if parabola opens up, or from the vertex down, if the parabola turns downward. Thanks to completing the square, you have a tool to rewrite the equation of the function in what is called *vertex form:* $y = a(x - h)^2 + k$. The vertex will be located at (h, k), and *a* will tell you whether the parabola opens up or down. This form also makes the transformations clear.

> 📖 **DEFINITION**
>
> **Vertex form** of the equation of a parabola, $y = a(x - h)^2 + k$, is achieved by completing the square on $y = ax^2 + bx + c$ and shows the coordinates of the vertex, or turning point, of the parabola (h, k).

To change from $y = ax^2 + bx + c$ form to $y = a(x - h)^2 + k$, you can use the completing the square technique. For instance, to put $y = 3x^2 + 12x + 7$ in vertex form:

1. Move the constant to the *y* side.

   $$y - 7 = 3x^2 + 12x$$

2. Divide through by 3.

   $$\frac{y-7}{3} = x^2 + 4x$$

3. Complete the square.

   $$\frac{y-7}{3} + 2^2 = x^2 + 4x + 2^2$$

4. Write as a square.

$$\frac{y-7}{3}+4=\left(x+2\right)^2$$
$$\frac{y-7}{3}+\frac{12}{3}=\left(x+2\right)^2$$
$$\frac{y+5}{3}=\left(x+2\right)^2$$

5. Isolate y.

$$\frac{y+5}{3}=\left(x+2\right)^2$$
$$y+5=3\left(x+2\right)^2$$
$$y=3\left(x+2\right)^2-5$$

The vertex of the parabola is at $(-2, -5)$, and the parabola opens up, with a vertical stretch.

Now that you know how to put the equations in vertex form, let's try graphing some. To graph $f(x) = -(x - 3)^2 - 4$, shift the parent graph 3 units right, reflect it over the x-axis, and move down 4 units.

Key Points	3 Right	Reflect	4 Down
(0, 0)	(3, 0)	(3, 0)	(3, −4)
(2, 4)	(5, 4)	(5, −4)	(5, −8)
(−2, 4)	(1, 4)	(1, −4)	(1, −8)

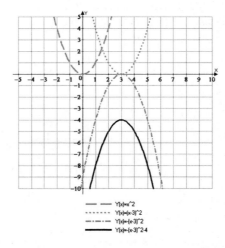

To graph $g(x) = 3(x + 2)^2 - 5$, shift 2 units left, stretch vertically by a factor of 3, and then shift down 5.

Key Points	2 Left	Stretch	5 Down
(0, 0)	(–2, 0)	(–2, 0)	(–2, –5)
(2, 4)	(0, 4)	(0, 12)	(0, 7)
(–2, 4)	(–4, 4)	(–4, 12)	(–4, 7)

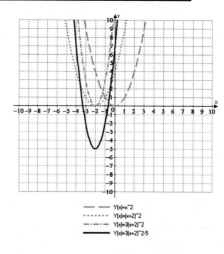

Y(x)=x^2
Y(x)=(x+2)^2
Y(x)=3(x+2)^2
Y(x)=3(x+2)^2-5

ALGEBRA TIP

The graph of the parent parabola $y = x^2$ shows a pattern of increase in the y-coordinates by odd numbers. On each side of the vertex, the y-coordinates are 1, 4, 9, 16, and so on. The increase from one to the next is 1 unit, and then 3, and then 5, and so on. You can use this pattern, multiplied by the coefficient a, to sketch the parabola quickly.

CHECKPOINT

Sketch a graph of each function.

26. $f(x) = (x - 4)^2 + 3$

27. $y = 2(x + 5)^2 - 3$

28. $g(x) = -3(x + 1)^2 - 5$

29. $y = -\frac{1}{2}(x - 4)^2 + 1$

30. $f(x) = 4(x + 3)^2 - 7$

Quadratic Modeling

There are situations in which you collect data, or are given data, about the relationship between two variables. You might collect data about annual income and amount of federal income tax paid, for example. Many times, you graph that data and look for the line of best fit.

When data collected doesn't suggest a line, but rather has a curvature that suggests a parabola, you may want to fit a quadratic equation to the data. Instead of starting with the equation and finding the parabola, you're starting with what looks like a parabola and trying to find a quadratic function that describes it. Technology can be helpful, but if that's not available, you can choose three points that represent the data set well and use them to find the equation of a parabola that might describe the data.

If one point is the vertex of the parabola, you can start with vertex form and substitute the coordinates of the vertex for h and k and the coordinates of one other point for x and y. That will allow you to solve to find the value of a, and you'll have enough information to write the equation.

For instance, if you know the vertex of a parabola is located at $(5, -7)$ and that the points $(1, 25)$ and $(7, 1)$ are on the parabola, you can start with $y = a(x - 5)^2 - 7$ and plug one of the points in for x and y.

$$25 = a(1 - 5)^2 - 7$$

Simplify and solve for a.

$$25 = a(-4)^2 - 7$$
$$25 = 16a - 7$$
$$32 = 16a$$
$$a = 2$$

Now you can write the equation as $y = 2(x - 5)^2 - 7$.

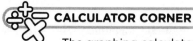 **CALCULATOR CORNER**

The graphing calculator is likely to have a function that will fit a quadratic equation to data. (In fact, most can fit several different types of functions to data. Check the menu.) The calculator does the calculation quickly, but it's up to you to recognize the shape of the graph and select the correct type.

If you're not certain where the vertex is, you can still find an equation, but it will take a bit more work. Suppose you have the points (–5, –5), (–2, –2), and (0, –10). Each point should fit an equation of the form $y = ax^2 + bx + c$, but you need to fit the correct values of a, b, and c.

Create a system of three equations, each of the form $y = ax^2 + bx + c$, but substitute $x = –5$ and $y = –5$ into the first one, $x = –2$ and $y = –2$ into the second, and $x = 0$ and $y = –10$ into the third.

$$-5 = a(-5)^2 + b(-5) + c$$
$$-2 = a(-2)^2 + b(-2) + c$$
$$-10 = a(0)^2 + b(0) + c$$

Simplify and you'll find yourself with a system of equations.

$$-5 = 25a - 5b + c$$
$$-2 = 4a - 2b + c$$
$$-10 = c$$

In this case, the third equation becomes trivial, but even when that doesn't happen, you can solve a three variable system. In this case, you can substitute –10 for c in the other equations, and you'll have a two-variable system.

$$\begin{array}{cc} -5 = 25a - 5b - 10 & \quad 25a - 5b = 5 \\ -2 = 4a - 2b - 10 & \rightarrow \quad 4a - 2b = 8 \end{array}$$

$$50a - 10b = 10$$
$$\underline{-20a + 10b = -40}$$
$$30a = -30$$
$$a = -1$$
$$4(-1) - 2b = 8$$
$$-2b = 12$$
$$b = -6$$

Once you know that $a = –1$, $b = –6$, and $c = –10$, you can write the equation of the parabola: $y = -x^2 - 6x - 10$.

CHECKPOINT

Find the equation of a quadratic function that passes through the given points.

31. Vertex (–3, 7) and passing through (1, –41)

32. Vertex (4, –1) and passing through (2, 1)

33. Vertex (–5, –4) and passing through (–2, 23)

34. Passing through the points (–2, 17), (1, –1), and (6, 9)

35. Passing through the points (–2, –18), (–1, –5), and (2, 10)

The Least You Need to Know

- Solve quadratic equations by the square root method, by completing the square, or by the quadratic formula (and of course, by factoring).

- Use the discriminant to determine the number and type of solutions.

- Complex numbers, with real part a and imaginary part bi, allow you to solve equations with nonreal solutions. Remember, $i = \sqrt{-1}$.

- Graph quadratic functions by transforming the parent parabola.

- Find the equation of a parabola by plugging known points into vertex form, $y = a(x - h)^2 + k$, or create a system of equations to find the values of the coefficient of $y = ax^2 + bx + c$.

Conics

Quadratic functions have a predictable graph—a parabola opening either up or down—and a domain of all real numbers. There are, however, other quadratics—quadratic relations—that present in a greater variety of ways. In this chapter, you look at some of those quadratic relations, their principal characteristics, and methods for quickly sketching their graphs. You also see how methods you used to solve linear systems can be applied to solve systems with some of these quadratics.

In This Chapter

* Recognizing the conic sections
* Putting the equation of a conic in standard form
* Sketching the graph of each conic section
* Finding the equation of a conic from its graph
* Solving a system of quadratic equations

Parabolas

You've already graphed *parabolas* that open up or down, but what if you were to rotate one of those parabolas 90°, as shown in the following figure?

Suddenly, that parabola is no longer a function. It fails the vertical line test at every point except the vertex. Instead of a domain of all real numbers and a range of $y \geq 0$, this rotated parabola has a domain of $x \geq 0$ and a range of all real numbers. If you feel like the x's and y's have changed places, you're not far off. The equation of this parabola that opens to the right and has its vertex at the origin is $x = y^2$.

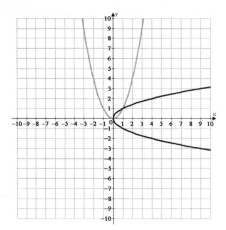

This is one of a group of quadratic relations called *conic sections,* because the shapes of their graphs are cross-sections of a cone sliced by a plane. Which shape is produced depends on the angle of the plane. The parabola is produced when the plane cuts from the side to the base of the cone at an angle between vertical and horizontal, which is shown as follows.

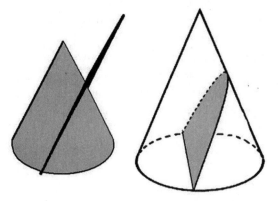

The parabola is defined as the set of all points in the plane whose distance from a point called the *focus* and a line called the *directrix* is constant. The vertex or turning point of the parabola sits on the imaginary line called the *axis of symmetry.* The two sides of the parabola are reflections of one another across the axis of symmetry. The focus also lies on the axis of symmetry, and its distance from the vertex is called the *focal length.*

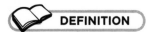 **DEFINITION**

A **parabola** is a conic section described as the set of points equidistant from a point called the *focus* and a line called the *directrix*. A **conic section** is the intersection of a plane and a cone. The intersection forms a figure that can be described by a quadratic relation.

On the opposite side of the vertex from the focus, at the same distance, is the directrix, which is perpendicular to the axis of symmetry. If you choose any point on the parabola and measure the perpendicular distance from that point to the directrix, and the distance from that point on the parabola to the focus, those distances will be the same, as you can see in the following graph.

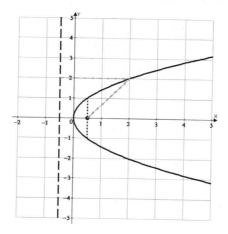

Vertex Form Equation

The vertex form of the equation of a parabola that opens up or down is $y = a(x - h)^2 + k$ or $y - k = a(x - h)^2$. When the parabola opens to the right or left, the $x - h$ and $y - k$ change places, and the vertex form is $x - h = a(y - k)^2$. In both cases, the vertex is (h, k).

To sketch the graph of a parabola quickly, you want the equation in vertex form. If the given equation is not in vertex form, you can use completing the square (see Chapter 8) to transform it. For example, the parabola with equation $x = -2y^2 - 4y + 1$ can be transformed to vertex form.

$$x = -2y^2 - 4y + 1$$
$$x - 1 = -2y^2 - 4y$$
$$\tfrac{x-1}{-2} = y^2 + 2y$$
$$\tfrac{x-1}{-2} + 1 = y^2 + 2y + 1$$
$$\tfrac{x-1}{-2} + \tfrac{-2}{-2} = (y+1)^2$$
$$\tfrac{x-3}{-2} = (y+1)^2$$
$$x - 3 = -2(y+1)^2$$

In this form, you can see that $x - h$ is $x - 3$, so the vertex has an x-coordinate of 3, and $y - k$ is $y + 1$, so $k = -1$. The vertex is $(3, -1)$ and the coefficient -2 tells you that parabola opens to the left. It also tells you the parent function has been stretched by a factor of 2.

CALCULATOR CORNER

While most graphing calculators are designed to graph functions, most conics are not functions. You may find it more convenient to sketch the graphs by hand, but if you want to create the graph on your calculator, isolate y. When you do, you'll have to take the square root of both sides, and that will result in both a positive and a negative square root. Enter two equations, one using the positive square root and one with the negative root. You may need to adjust the window dimensions to make the graph look correct.

CHECKPOINT

Put each equation in vertex form.

1. $y^2 + 8y + 4x + 24 = 0$

2. $8y - x^2 = 4x + 20$

3. $3y^2 + 9x = 6y + 24$

4. $5y - 10x^2 = 40x + 20$

5. $x + 7y^2 = 14y + 2$

Graphing a Parabola

Although the focus and directrix do not appear directly in the equation, the value of a—the coefficient that tells you about direction of opening and the stretch—is connected to the focal length. If p is the focal length and a is the coefficient, $4ap = 1$. The larger the focal length, the smaller the coefficient a and the wider the parabola appears. You can use the equation $4ap = 1$ to calculate the focal length. Using the example from the previous section, $4(-2)p = 1$, so $p = -\frac{1}{8}$. That places the focus at $\left(3 - \frac{1}{8}, -1\right) = \left(2\frac{7}{8}, -1\right)$. The directrix is the line $x = 3\frac{1}{8}$.

The *latus rectum* is a line segment parallel to the directrix and perpendicular to the axis of symmetry that passes through the focus and has its endpoints on the parabola. The length of the latus rectum is always $4p$, so it can help you set the width of the parabola. The latus rectum has a length of $4\left(-\frac{1}{8}\right) = -\frac{1}{2}$ (the sign is not significant here) and has the focus as its center. Its endpoints fall at $\left(2\frac{7}{8}, -1 + \frac{1}{4}\right) = \left(2\frac{7}{8}, -\frac{3}{4}\right)$ and $\left(2\frac{7}{8}, -1 - \frac{1}{4}\right) = \left(2\frac{7}{8}, -1\frac{1}{4}\right)$.

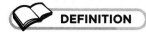 **DEFINITION**

> The **latus rectum** is a line segment parallel to the directrix of a parabola, which has the focus as its midpoint and whose endpoints rest on the parabola.

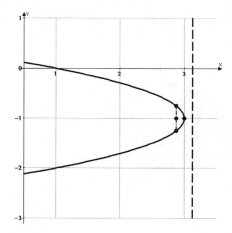

Plotting the vertex and the endpoints of the latus rectum will help you set the shape of the parabola. You can use the x-intercept and y-intercepts to complete the graph.

If you have sufficient information about the parabola, you can find the equation. For instance, the equation of a parabola with focus (6, 1) and directrix $x = 4$ can be found by first determining where the vertex falls. A sketch will help you determine direction of opening and therefore which version of the equation you want. The parabola always wraps around its focus. The vertex is midway between the focus and the directrix, so it must be (5, 1). The focal length is $p = 1$, so $4a(1) = 1$ and $a = \frac{1}{4}$. That's all you need to conclude that the equation of this parabola is $x - 5 = \frac{1}{4}(y - 1)^2$.

 CHECKPOINT

Sketch the graph of each parabola. Show the focus and directrix.

6. $y = \frac{1}{2}(x - 2)^2 + 5$

7. $x - 7 = -3(y + 1)^2$

8. $y = -5(x + 4)^2 - 7$

9. $x + 3 = -4(y - 3)^2$

10. $x - 5 = -\frac{1}{4}(y + 3)^2$

Circles

You're probably well acquainted with *circles* from geometry, but like the other quadratic relations, circles are conic sections, created when the plane slices the cone parallel to the base of the cone (which you can see in the following image). The circle is defined as the set of all points at a fixed distance called the *radius* from a fixed point called the *center*. The standard form of the equation of a circle with center (h, k) and radius r is $(x - h)^2 + (y - k)^2 = r^2$.

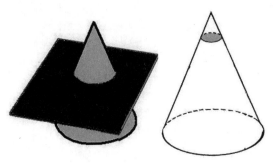

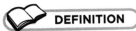 **DEFINITION**

A **circle** is the set of all points at a fixed distance called the *radius* from a center point.

To sketch the graph quickly, all you need is the center and the radius, both clear when the equation is in standard form. To graph $(x - 3)^2 + (y + 2)^2 = 16$, for example, find the center $(3, -2)$ and from that point, count 4 up, 4 down, 4 left, and 4 right. Finally, connect the dots with the smoothest curve you can manage.

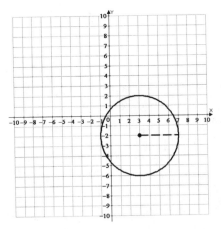

Putting the equation of a circle in standard form takes the same steps as the equation of an ellipse, so let's hold that discussion for a moment. However, to simply write the equation of a circle, all you need is the center and the radius to drop into place. So the circle with a center at (7, –2) and a radius of 9 has the equation $(x - 7)^2 + (y + 2)^2 = 81$.

CHECKPOINT

Sketch the graph of each circle.

11. $(x + 5)^2 + (y - 2)^2 = 4$ 14. $x^2 + (y + 5)^2 = 49$

12. $(x - 3)^2 + (y - 1)^2 = 25$ 15. $(x - 2)^2 + y^2 = 9$

13. $(x + 4)^2 + (y - 1)^2 = 16$

Ellipses

People tend to think of an *ellipse* as a squashed circle, because most of us meet the circle first, and so see the ellipse as derived from the circle. It's actually more the reverse: the circle is a special case of the ellipse. The shape called the ellipse is formed when the cone is sliced from side to side, as shown in the following figure. A cut that's parallel to the base results in a circle, but most such slices would not be parallel to the base, and instead would make the oval shape called an ellipse.

DEFINITION

An **ellipse** is an oval shape formed by all the points for which the sum of the distances from two focal points is constant.

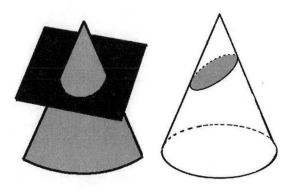

The ellipse is defined as the set of all points for which the sum of the distances from the point to two focal points is a constant. Only two points on the ellipse are equidistant from the two foci, but when a point is close to one focus, it is farther from the other, and the total of the two distances is always the same. The distance from the center to a focal point is the focal length, usually labeled as c. The foci are equidistant from the center, so the distance between the foci is $2c$.

The ellipse doesn't have a consistent radius. Instead, it has a longer major axis and a shorter minor axis. The axes cross at the center of the ellipse, and the foci always sit on the major axis. The ends of the major axis are called *vertices*, and the ends of the minor axis are *co-vertices*.

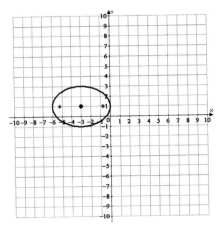

The distance from the center to a focal point is usually labeled as c. The length of the major axis is $2a$, and $2b$ is the length of the minor axis. The equation $c^2 = a^2 - b^2$ relates the focal length to the lengths of the axes in any ellipse.

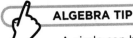

ALGEBRA TIP

A circle can be seen as an ellipse whose foci have converged at the center. If the focal length $c = 0$, $0^2 = a^2 - b^2$ suggests that the lengths of the major and minor axes are the same, producing a circle. There is no need to learn separate techniques for circles, because those for ellipses suffice.

Standard Form Equation

The standard form of the equation of an ellipse is $\frac{(x-h)^2}{a^2} + \frac{(y-k)^2}{b^2} = 1$ if the major axis is horizontal or $\frac{(x-h)^2}{b^2} + \frac{(y-k)^2}{a^2} = 1$ if the major axis is vertical. If the ellipse equation is not in standard form, you can use completing the square to transform it, but you'll need to do it twice—once for the x's and once for the y's.

For instance, the equation $9x^2 + 4y^2 + 54x - 40y + 145 = 0$ can be coaxed into standard form, but first, group the x-terms together, group the y-terms together, and move the constant to the other side.

$$(9x^2 + 54x) + (4y^2 - 40y) = -145$$

Focus on the x-terms first. Factor out the 9, and then complete the square for the expression inside the parentheses. When you add to the other side, remember that the constant you placed inside the parentheses will be multiplied by that common factor of 9, so add 9 times 9 or 81.

$$9\left(x^2 + 6x\right) + \left(4y^2 - 40y\right) = -145$$
$$9\left(x^2 + 6x + 9\right) + \left(4y^2 - 40y\right) = -145 + 9 \cdot 9$$
$$\left(x + 3\right)^2 + \left(4y^2 - 40y\right) = -64$$

Turn your attention to the y-terms. Factor out the common factor of 4 and complete the square. Remember to adjust for that 4 when you add to the other side.

$$9\left(x + 3\right)^2 + \left(4y^2 - 40y\right) = -64$$
$$9\left(x + 3\right)^2 + 4\left(y^2 - 10y\right) = -64$$
$$9\left(x + 3\right)^2 + 4\left(y^2 - 10y + 25\right) = -64 + 4 \cdot 25$$
$$9\left(x + 3\right)^2 + 4\left(y - 5\right)^2 = 36$$

Divide through by the constant—in this case, 36.

$$9\left(x + 3\right)^2 + 4\left(y - 5\right)^2 = 36$$
$$\frac{9(x+3)^2}{36} + \frac{4(y-5)^2}{36} = \frac{36}{36}$$
$$\frac{(x+3)^2}{4} + \frac{(y-5)^2}{9} = 1$$

CHECKPOINT

Put the equation of each ellipse in standard form.

16. $x^2 + y^2 + 4x + 6y - 3 = 0$

17. $121x^2 + 25y^2 - 150y = 2{,}800$

18. $3x^2 + 5y^2 - 36x + 30y + 138 = 0$

19. $2x^2 + 3y^2 + 16x - 30y + 101 = 0$

20. $4x^2 + 25y^2 + 8x - 50y - 71 = 0$

Graphing an Ellipse

To sketch the graph of an ellipse, you don't really need to worry about major axis or minor axis; that will work itself out. To graph $\frac{(x-1)^2}{49} + \frac{(y+4)^2}{16} = 1$, for instance, first locate the center. In this case, the center is $(1, -4)$. Under $(x - 1)^2$, you see 49. The denominator of the fraction with x in the numerator will tell you about the horizontal axis. Take the square root and count 7 left and 7 right from the center to mark the ends of the axis. For the vertical axis, take the square root of 16 and count up 4 and down 4 from the center. This should give you enough to sketch the ellipse. You can see that the horizontal axis is longer, so that's the major axis.

The foci are left and right of center, at a distance of $c = \sqrt{a^2 - b^2} = \sqrt{49 - 16} = \sqrt{33} \approx 5.75$. That places them at approximately $(1 - 5.75, -4)$ and $(1 + 5.75, -4)$, or $(-4.75, -4)$ and $(6.75, -4)$, which you can see in the figure.

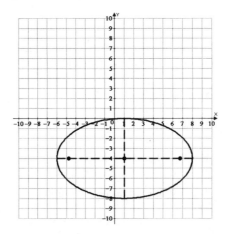

To write the equation of an ellipse, you'll need the center and enough information to determine the lengths of the major and minor axes. In other words, you need the center and the values of a and b. You may have those values directly, if you're actually looking at the graph, or you may have to find one or more of them from other information.

For example, to find the equation of an ellipse centered at $(1, 2)$ with a focus at $(1, 5)$ and passing through the point $(-3, 2)$, you'll want to start by plotting the points to get yourself oriented.

You can see that the focus is on the vertical axis, so that is the major axis, and the other focal point should be 3 units below the center at $(1, -1)$. The point $(-3, 2)$ is directly in line with the center, so it will be the end of the minor axis, and the other end will be $(5, 2)$, as shown in the following graph.

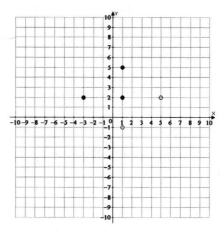

You can count and see that $c = 3$ and $b = 4$. You then can calculate a by using $c^2 = a^2 - b^2$.

$$c^2 = a^2 - b^2$$

$$3^2 = a^2 - 4^2$$

$$9 + 16 = a^2$$

$$a^2 = 25$$

$$a = 5$$

You have the center and the values of a and b, so you can write the equation $\frac{(x-1)^2}{4^2} + \frac{(y-2)^2}{5^2} = 1$ or $\frac{(x-1)^2}{16} + \frac{(y-2)^2}{25} = 1$. You can also complete the graph, if you wish, as follows.

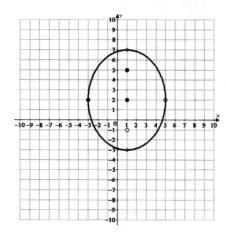

CHECKPOINT

Sketch the graph of each ellipse. Mark the foci.

21. $\frac{(x-3)^2}{16}+\frac{(y-2)^2}{25}=1$ 22. $\frac{(x-4)^2}{9}+\frac{y^2}{4}=1$

23. $\frac{(x-1)^2}{16}+\frac{(y-2)^2}{4}=1$ 24. $(x+5)^2+\frac{(y-1)^2}{4}=1$

25. Sketch a graph and find an equation for the ellipse with a center at (0, 0), focus at (–3, 0), and vertex at (5, 0).

Hyperbolas

The last conic is created when a plane slices a double-napped cone vertically. The cross-section is two curves turning away from one another, as shown in the following figure. The definition of a *hyperbola* sounds very similar to the definition of an ellipse, but the hyperbola is the set of all points for which the difference (not the sum) of the distances from two foci is constant. Each of the two curves that make up the hyperbola wraps around one of the foci.

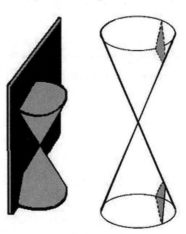

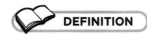

DEFINITION

The **hyperbola** is made up of a pair of curves, reflections of one another, which curve away from one another around two focal points. The distances between a point of the hyperbola and each of the two focal points subtract to a constant value.

Because the definition of the hyperbola is so similar to the definition of the ellipse, it's not surprising that their equations are similar as well. Like the ellipse, the hyperbola has a center, but the way you think about the axes is a little different. The axis that the curves intersect is called the *transverse axis,* but there is no distinction of major or minor axis. The transverse axis may be longer, shorter, or the same length as the other axis.

Standard Form Equation

The standard form of the equation of the hyperbola contains an $(x - h)^2$ and a $(y - k)^2$ that show the center (h, k). If the transverse axis is horizontal, the equation is $\frac{(x-h)^2}{a^2} - \frac{(y-k)^2}{b^2} = 1$. If the transverse axis is vertical, the equation is $\frac{(y-k)^2}{b^2} - \frac{(x-h)^2}{a^2} = 1$. When it comes to the equation of a hyperbola, remember these rules:

- The order of the subtraction tells you whether the transverse axis is horizontal $(x^2 - y^2)$ or vertical $(y^2 - x^2)$.

- The center is (h, k), but make sure you get the order correct. Look for h in $x - h$ and k in $y - k$. Don't just read left to right.

- The denominator under the x's tells you how to count left and right, and the one under the y's tells you how to count up and down.

The good news is that putting the equation of a hyperbola in standard form is basically the same process as putting an ellipse in standard form. For example, to put $4y^2 - x^2 + 6x + 32y + 39 = 0$ into standard form, group terms with x's and terms with y's and move the constant to the other side.

$$4y^2 - x^2 + 6x + 32y + 39 = 0$$
$$4y^2 + 32y - x^2 + 6x = -39$$

Factor 4 out of the y-terms and -1 out of the x-terms.

$$4(y^2 + 8y) - 1(x^2 - 6x) = -39$$

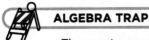

ALGEBRA TRAP

The most common errors that occur when putting the equation of a hyperbola in standard form are those that arise from not handling the minus sign between the x- and y-terms correctly. Group the terms for each variable, and carefully factor a negative common factor out of all the terms in the group that begins with a negative square term. When completing the square, remember to multiply the new constant by that negative factor before adding to the other side.

Complete the square, remembering to adjust the constant for the greatest common factor (GCF) of 4 and the GCF of -1.

$$4\left(y^2 + 8y + 16\right) - 1\left(x^2 - 6x + 9\right) = -39 + 4\cdot16 - 1\cdot9$$

$$4\left(y + 4\right)^2 - \left(x - 3\right)^2 = 16$$

Divide through by 16.

$$\frac{4(y+4)^2}{16} - \frac{(x-3)^2}{16} = \frac{16}{16}$$

$$\frac{(y+4)^2}{4} - \frac{(x-3)^2}{16} = 1$$

CHECKPOINT

Put the equation of each hyperbola in standard form.

26. $x^2 - y^2 - 2x - 4y = 28$

27. $9x^2 + 18x - 4y^2 - 16y = 43$

28. $16y^2 + 18x - 89 = 9x^2 + 64y$

29. $x^2 - 25y^2 + 4x - 200y - 421 = 0$

30. $25x^2 - y^2 + 4y - 4 = 25$

Graphing a Hyperbola

Graphing hyperbolas shares much with graphing ellipses. You'll locate the center as a place to start, and then count left and right and up and down to mark vertices. Unlike the ellipse, which is bounded by those points, the hyperbola flies out in two curves, so you'll need to take an extra step to find guide lines for those curves. Let's look at an example.

To sketch the graph of the hyperbola for the example in the previous section, start at the center $(3, -4)$. The 16 tells you to count 4 to the left and 4 to the right, and the 4 says the vertical axis goes 2 up and 2 down. Put a dot at each of those points, and lightly sketch a rectangle whose sides pass through those points. You'll use that rectangle to sketch the asymptotes of the graph, as you can see in the following.

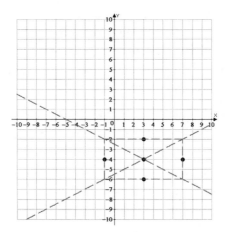

The asymptotes are lines that pass through the center and have slopes of $\pm\frac{b}{a}$. The curves approach these asymptotes closely as they move away from the center. When you sketch the graph, these asymptotes will be the diagonals of the rectangle, extended. Place the curves so they intersect the transverse axis at the edge of the rectangle and approach the asymptotes as they move outward, as shown in the following graph.

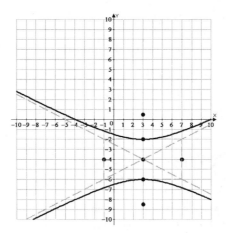

ALGEBRA TIP

Remember that the asymptotes are not actually part of the graph, just as the center of the circle is not a point of the circle and the directrix is not part of the parabola. Drawing the asymptotes is a great help in sketching the graph, but only the two curves are part of the hyperbola.

The foci for this hyperbola can be found using the equation $c^2 = a^2 + b^2$. You have $a = 4$ and $b = 2$, so $c^2 = 4^2 + 2^2 = 16 + 4 = 20$. Take the square root and $c = 2\sqrt{5} \approx 4.47$. This means the foci are on the transverse axis at $(3, -4 \pm 2\sqrt{5})$, or approximately $(3, -8.47)$ and $(3, 0.47)$.

Finding the equation of a hyperbola follows the same pattern as finding the equation of an ellipse. You need to know the center and you need enough information to find the values of a and b. You may be given those by being told the lengths of axes, or you may be given the foci and a vertex, and have to use $c^2 = a^2 + b^2$ to find the missing one. Once you have a and b, you substitute those values and the center into the appropriate form of the equation.

For example, suppose you know that a certain hyperbola is centered at $(-7, 4)$, and has a vertex at $(-7, 7)$ and a focus at $(-7, 8)$. If you plot those points, you can see that the hyperbola will open up and down, so its equation has the form $\frac{(y-k)^2}{a^2} - \frac{(x-h)^2}{b^2} = 1$. The center, $(-7, 4)$ goes in for (h, k) and you have $\frac{(y-4)^2}{a^2} - \frac{(x+7)^2}{b^2} = 1$. Now you need a and b. The vertex is 3 units above the center, so $a = 3$, and the focus is 4 units above the center, so $c = 5$. Using $c^2 = a^2 + b^2$, you find that $b^2 = 25 - 16$ and $b = 4$. Put a and b in their places, and you have the equation $\frac{(y-4)^2}{3^2} - \frac{(x+7)^2}{4^2} = 1$ or $\frac{(y-4)^2}{9} - \frac{(x+7)^2}{16} = 1$.

CHECKPOINT

Sketch the graph of each hyperbola. Show the asymptotes and mark the foci.

31. $\frac{y^2}{4} - \frac{x^2}{36} = 1$ 32. $\frac{(y-2)^2}{36} - \frac{(x+2)^2}{4} = 1$ 33. $\frac{(x+5)^2}{9} - \frac{(y-3)^2}{25} = 1$

34. $(x - 2)^2 - 4(y - 1)^2 = 4$

35. Sketch a graph and find an equation for the hyperbola that has vertices at $(0, \pm2)$, if its asymptotes are the lines $y = \pm\frac{2}{3}x$.

Quadratic Systems

Your experience with solving systems of equations has focused on solving linear systems, or systems in which all the equations had graphs that were lines. Now you have the tools to tackle *nonlinear systems*—systems in which one or more of the equation have graphs that curve. Although *nonlinear* could refer to any graph that isn't a line, nonlinear systems usually involve second-degree equations, like the conics of this chapter. As a result, they're often called *quadratic systems.*

Systems of equations that involve one or more quadratic equations can look intimidating, but many of them can be solved with techniques you know well from solving linear systems: graphing, substitution, and elimination. Although the presence of squared terms makes the system look complicated, realize that an equation with one squared term is a parabola, and equations that contain both an x^2 and a y^2 will be circles, ellipses, or hyperbolas. You have the techniques to sketch those graphs. You may not be able to read the solution directly from the graph, because the intersection(s) may not occur at integer points. Those graphs will tell you how many intersection points to expect, however, and approximately what the points are. Substitution or elimination will give you the exact values.

 **DEFINITION**

> A **nonlinear system** is a system of equations that contains at least one equation of degree 2 or higher. Because of the variety of curves possible, you may find no solution or one, two, or more solutions.

Sketching the graph of the equations in the system allows you to anticipate how many solutions the system will have, even if it is not possible to read those solutions from the graph. If you can isolate one of the variables or the square of one of the variables in one equation, you can substitute into the other equation. You'll probably have a quadratic in one variable to solve. Remember to substitute back in to find the value of the other variable, as you learned in Chapter 5.

Solving the following system begins with a sketch. The first equation gives you a parabola that opens to the right. The second equation is an ellipse, centered at the origin.

$$x = y^2 - 4$$
$$x^2 + 4y^2 = 36$$

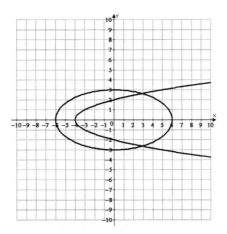

You can see from the graph that there will be two points of intersection, and they don't appear to have integer coordinates. The first equation looks simple enough to make substitution possible, but if you replace x with $y^2 - 4$ you'll end up with a fourth power. Instead, isolate $y^2 = x + 4$, and replace y^2 with $x + 4$. That substitution gives you a more manageable quadratic. It will still require the quadratic formula, however, and you should be alert for extraneous solutions.

$$x = y^2 - 4 \rightarrow y^2 = x + 4$$
$$x^2 + 4y^2 = 36$$
$$x^2 + 4(x + 4) = 36$$
$$x^2 + 4x + 16 = 36$$
$$x^2 + 4x - 20 = 0$$
$$x = \frac{-4 \pm \sqrt{16+80}}{2} = \frac{-4 \pm \sqrt{96}}{2} = \frac{-4 \pm 4\sqrt{6}}{2} = -2 \pm 2\sqrt{6}$$
$$x \approx 2.9 \text{ or } x \approx -6.9$$

Your graph should tell you that −6.9 is extraneous. Neither equation is defined for any $x < -6$. Plug in 2.9 for x to find the corresponding values of y. Based on this, the solutions of the system are approximately (2.9, 2.6) and (2.9, −2.6).

Systems that don't lend themselves to substitution may yield to elimination. The ellipse $4x^2 + 9y^2 = 36$ and the circle $x^2 + y^2 = 6$ intersect in four points, as shown in the following graph.

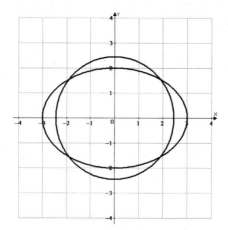

The system can be solved by multiplying the equation of the circle by −9 and adding the equations.

$$4x^2 + 9y^2 = 36$$
$$-9(x^2 + y^2) = (6)(-9)$$
$$\overline{}$$

$$4x^2 + 9y^2 = 36$$
$$-9x^2 - 9y^2 = -54$$
$$\overline{}$$
$$-5x^2 = -18$$
$$x^2 = 3.6$$
$$x \approx \pm 1.9$$

$$x^2 + y^2 = 6$$
$$3.6 + y^2 = 6$$
$$y^2 = 2.4$$
$$y \approx \pm 1.5$$

CHECKPOINT

Solve each system.

36. $x^2 + 4y^2 = 16$
    $x + 2y = -4$

37. $x^2 + y^2 = 9$
    $4x^2 + 25y^2 = 100$

38. $25(x-3)^2 + 16(y-2)^2 = 400$
    $(x-3)^2 + (y+1)^2 = 4$

39. $x^2 - y^2 - 2x - 4y = 28$
    $x^2 + y^2 + 4x + 4y = 56$

40. $y^2 + 8x + 4y + 24 = 0$
    $x^2 + y^2 + 4y + 6x - 3 = 0$

The Least You Need to Know

- Sketch the graph of a parabola from $y - k = a(x - h)^2$ or $x - h = a(y - k)^2$ by plotting the vertex (h, k) and using a to set the width and direction of opening. Use $4ap = 1$ to find the focal length p. The directrix is a line p units to the other side of the vertex, perpendicular to the axis of symmetry.

- Sketch the graph of an ellipse $\frac{(x-h)^2}{a^2} + \frac{(y-k)^2}{b^2} = 1$ or circle $(x - h)^2 + (y - k)^2 = r^2$ by locating the center (h, k). You can then use a and b to count out from the center to the ends of the major and minor axes of the ellipse, or r to count in all four directions for the circle. Find the distance c, from center to foci, with the equation $c^2 = a^2 - b^2$.

- Sketch the graph of a hyperbola by locating the center (h, k) and then using a and b to count out from the center. Draw a rectangle whose sides pass through those four points, and extend the diagonals that serve as asymptotes. The equations of the asymptotes are $y = \pm\frac{b}{a}(x - h) + k$.

- If the equation of a hyperbola is $\frac{(x-h)^2}{a^2} - \frac{(y-k)^2}{b^2} = 1$, the hyperbola's two curves touch the left and right sides of the rectangle. If the equation is $\frac{(y-k)^2}{a^2} - \frac{(x-h)^2}{b^2} = 1$, the curves touch the top and bottom.

- Solve systems involving equations of conics by using one equation to substitute into the other, either for a variable or for its square. Alternately, add or subtract multiples of the equations to eliminate one variable.

PART 4

Polynomial and Rational Functions

Ready for more questions? I don't mean problems, in this case, but the kind of questions you might already be asking—the wonderings of an interested student. If there's a quadratic formula, is there a cubic formula? Yes, but sadly not one you'd want to memorize. How do you deal with equations involving terms of a degree higher than 2? What happens when you build fractions from polynomials?

Answering some of those questions is the task of this part. You examine a set of techniques that will let you search for the solutions of a polynomial of any degree, a task made much simpler these days thanks to graphing calculators. You learn what's predictable and what's less predictable about the graphs of polynomial functions. Before you move on, you build a whole system of arithmetic for those algebraic fractions, and acquire a few handy methods for simplifying, solving, and graphing.

Polynomials

You began your study of algebra with linear equations, and then stepped up to quadratics. That increase of one degree introduced new complexities in graphing and new techniques for solving. But what if you increase the degree even more? In this chapter, you look at functions of a degree greater than 2, and explore their graphs and methods for solving them.

In This Chapter

- Reducing polynomials to a product of factors
- Using long division and synthetic division
- Solving polynomial inequalities
- Graphing polynomial functions

Polynomial Equations and Functions

A *polynomial* is defined as a sum of terms, each of which is a product of a real number and a variable raised to a power. The exponents must be non-negative integers. The definition is often given by showing the pattern $a_n x^n + a_{n-1} x^{n-1} + \ldots + a_2 x^2 + a_1 x + a_0$. The subscripts on the a's simply relate them to a particular term. For instance, a_5 would mean "the coefficient belonging to the fifth power term."

 **DEFINITION**

A **polynomial** is a sum of terms, each of which is a product of a real number and a non-negative integer power of a variable.

Viewed as a function, every polynomial has a domain of all real numbers. The degree of the polynomial is the highest power of the terms. For polynomials of odd degree, the range is all real numbers, but in polynomials of even degree, the range is restricted. You've seen this already in linear equations, which are first-degree polynomials with a range of all reals, and quadratic equations, which are second-degree polynomials with a range from the vertex up or down.

Solving polynomial equations borrows some key ideas from solving quadratics, which is not surprising because quadratics are polynomials. You have several methods for solving quadratics, and your principal tactic for solving higher-degree equations will be to find your way back to a quadratic.

When you think about a polynomial equation like $x^4 - 5x^2 + 6 = 0$, you talk about the solutions of the equation. When you think about the graph of $y = x^4 - 5x^2 + 6$, you think about its x-intercepts. When you consider the polynomial function $f(x) = x^4 - 5x^2 + 6$, you refer to the zeros of the function. Those terms—solution, x-intercept, and zero—are, if not perfect synonyms, almost the same thing, and are all connected to and derived from the factors of $x^4 - 5x^2 + 6$.

Although there are subtle differences in meaning, the zeros of a function, the roots or the solutions of an equation, and the x-intercepts of a graph all refer to the same numbers, and all trace back to the factors of the polynomial. Setting each factor equal to 0 and solving produces the values.

Factors and Solutions

One of your crucial techniques for solving quadratics is factoring. The zero-product property tells you that if you can re-express your quadratic as a product of factors equal to 0, then at least one of those factors must equal 0. That lets you translate the quadratic equation into two little linear equations that are easily solved. While that strategy is a great one for higher-degree equations as well, the challenge is finding the factors.

There are a few factoring patterns for cubic, or third-degree, polynomials that can be memorized, as you memorized the patterns for the difference of squares and the perfect square trinomial. Here are the patterns, with examples.

Sum of Cubes: $a^3 + b^3 = (a + b)(a^2 - ab + b^2)$

$8x^3 + 27 = (2x)^3 + 33 = (2x + 3)(4x^2 - 6x + 9)$

Difference of Cubes: $a^3 - b^3 = (a - b)(a^2 + ab + b^2)$

$125y^3 - 164 = (5y)^3 - 43 \ (5y - 4)(25y^2 + 20y + 16)$

Perfect Cube: $a^3 + 3a^2b + 3ab^2 + b^3 = (a + b)^3$

$x^3 + 6x^2 + 12x + 8 = x^3 + 3x^2 \cdot 2^2 + 3x \cdot 2^2 + 2^3 = (x + 2)^3$

While those patterns are great to know, they'll only take you so far. They only tell you how to factor cubics, and only some cubics. You're going to need a more general strategy, and that's going to be a bit of educated guessing—an informed trial-and-error. You're going to look for likely factors and try them out.

Two questions jump out immediately: how do you know what to try, and how do you try it? There are a few theorems that will help answer the first question.

CHECKPOINT

Factor each expression.

1. $t^3 + 30t^2 + 300t + 1{,}000$ 4. $125y^3 + 64$

2. $8x^3 - 125$ 5. $27x^3 - 8$

3. $z^3 + 1$

Rational Zeros Theorem

The theorem that lets you get started solving is the rational zeros theorem. It says that any solutions of the polynomial equation that are rational numbers will be the ratio of a factor of the constant term to a factor of the lead coefficient. Think about solving the linear equation $3x + 5 = 0$, for example. The solution is $-\frac{5}{3}$, and $\frac{5}{3}$ is the ratio of the constant to the lead coefficient.

The rational zeros theorem lets you make an initial list of possibilities that you might try. For instance, to find the possible rational zeros for the polynomial function $f(x) = 3x^3 - 10x^2 - 7x + 30$, start by listing the factors of the constant 30: 1, 2, 3, 5, 6, 10, 15, and 30. Next, make a list of the factors of the lead coefficient 3: 1 and 3. Based on these factors, the possible rational zeros are $\pm\frac{1}{1}, \pm\frac{2}{1}, \pm\frac{3}{1}, \pm\frac{5}{1}, \pm\frac{6}{1}, \pm\frac{10}{1}, \pm\frac{15}{1}, \pm\frac{30}{1}, \pm\frac{1}{3}, \pm\frac{2}{3}, \pm\frac{3}{3}, \pm\frac{5}{3}, \pm\frac{6}{3}, \pm\frac{10}{3}, \pm\frac{15}{3}, \pm\frac{30}{3}$.

Now, some of these are duplicates, so you can eliminate any occurrence after the first.

$$\pm\tfrac{1}{1}, \pm\tfrac{2}{1}, \pm\tfrac{3}{1}, \pm\tfrac{5}{1}, \pm\tfrac{6}{1}, \pm\tfrac{10}{1}, \pm\tfrac{15}{1}, \pm\tfrac{30}{1}, \pm\tfrac{1}{3}, \pm\tfrac{2}{3}, \cancel{\pm\tfrac{3}{3}}, \pm\tfrac{5}{3}, \cancel{\pm\tfrac{6}{3}}, \pm\tfrac{10}{3}, \cancel{\pm\tfrac{15}{3}}, \cancel{\pm\tfrac{30}{3}}$$

That's still a rather long list, and there was a time when you would have had little choice except to start trying them. But if you have access to a graphing calculator or other graphing software, you can look at the x-intercepts of the graph to see which numbers on the list are most promising.

CALCULATOR CORNER

Use a graphing utility to examine the x-intercepts of the graph of the polynomial function. Match intercepts to your list of possible rational zeros to narrow the list. Noninteger intercepts may or may not match the noninteger values on your list, but remember that the decimal value the trace feature gives you may have been rounded and so may not be a perfect match. It's better to test with synthetic division unless you're certain.

CHECKPOINT

List the possible rational zeros of each polynomial.

6. $f(x) = 2x^3 + 5x^2 + 3x - 8$ 9. $y = 6x^4 - 9x^3 + x + 14$

7. $y = 4x^3 - 9x^2 + 3x + 20$ 10. $f(x) = 9x^3 - 4x^2 + 2x - 15$

8. $g(x) = 6x^5 - 4x^3 - 5x - 4$

Descartes' Rule of Signs

In addition to help from the graphing utilities, you can narrow the list a bit by applying Descartes' rule of signs. According to Descartes' rule, the number of sign changes in a polynomial function corresponds to the maximum number of real zeros. To start, look at the polynomial—in this case, $f(x) = 3x^3 - 10x^2 - 7x + 30$—and count the number of times the sign changes as you move from term to term. For $(x) = 3x^3 - 10x^2 - 7x + 30$, the first term, $3x^3$, is positive. From $3x^3$ to $-10x^2$ is one sign change, but $-10x^2$ and $-7x$ are both negative, so that's still only one sign change. $-7x$ to $+30$ then gives you a second sign change.

Descartes' rule says that two sign changes means there are at most two positive real zeros. To find the maximum number of negative real zeros, replace x with $-x$, and count the sign changes in the resulting polynomial.

$$f(x) = 3(-x)^3 - 10(-x)^2 - 7(-x) + 30 = -3x^3 - 10x^2 + 7x + 30$$

This polynomial has only one sign change, which means the original, $f(x) = 3x^3 - 10x^2 - 7x + 30$, has a maximum of one negative real zero. The chances of a positive zero are better, so you'll probably want to start trying positive numbers.

ALGEBRA TIP

Evaluating $f(-x)$ for a polynomial function is as simple as changing the signs of the terms with odd powers.

CHECKPOINT

Determine the maximum number of positive real zeros and the maximum number of negative real zeros for each polynomial.

11. $f(x) = 2x^3 + 5x^2 + 3x - 8$ 14. $g(x) = 6x^4 - 9x^3 + x^2 + 14$

12. $g(x) = 4x^3 - 9x^2 + 3x + 20$ 15. $f(x) = 9x^3 - 4x^2 + 2x - 15$

13. $f(x) = 6x^5 - 4x^3 - 5x - 4$

Finding the Solutions

How will you know when you're done? Well, obviously, when you've found all the solutions. But how many should you expect? When you worked with quadratics, you had the discriminant to tell you whether to expect two real solutions; one real solution; or no real solutions, which meant two nonreal solutions. What about higher-degree functions?

The fundamental theorem of algebra tells you that over the complex numbers, a polynomial of degree n has n zeros. The number of zeros is equal to the degree of the polynomial, if you include the nonreal zeros. For example, the polynomial $f(x) = 3x^3 - 10x^2 - 7x + 30$ is a third-degree polynomial, so it will have three zeros. You're not guaranteed that all three will be real, though, so more information would be useful.

The key fact that will help here is that, for a polynomial function with real coefficients, if the complex number $a + bi$ is a zero, its conjugate, $a - bi$, is also a zero. Nonreal zeros will always come in conjugate pairs. If $f(x) = 3x^3 - 10x^2 - 7x + 30$ has three zeros, either all three are real, or one is real and there is a conjugate pair of complex zeros.

So even without the graphing calculator, you have some expectations about what your solutions may look like. However, there is one complication that I haven't addressed yet. The possible rational zeros are just that—rational. There may be real zeros that are irrational, and so don't appear on that list. You can't really guess at those, but if you can make your way to a quadratic factor, the quadratic formula will give you the exact value of those irrational zeros.

Now that you have some possibilities, how do you try them? Each possible zero comes from a factor. If you think 3 is a zero of the function (a solution of the equation $3x^3 - 10x^2 - 7x + 30 = 0$) then $x - 3$ will be a factor. Two theorems define the test for a factor:

- **Remainder theorem:** This says that if a function $f(x)$ is divided by $x - c$, the remainder is $f(c)$, the value you'd get if you plugged c in for x.

- **Factor theorem:** This says that if the remainder is 0, $x - c$ is a factor of the polynomial.

So if $3x^3 - 10x^2 - 7x + 30$ is divided by $x - 3$ and the remainder is 0, then $x - 3$ is a factor. There are two ways to accomplish this division.

Long Division

Polynomial *long division* is similar to the long division you learned in arithmetic. Using the example from the previous section, place the *dividend*, $3x^3 - 10x^2 - 7x + 30$, inside the division sign, and the divisor, $x - 3$, outside. Make your first estimate by dividing the first term of the dividend by the first term of the divisor: $\frac{3x^3}{x} = 3x^2$.

$$\boxed{x} - 3\overline{\smash{\big)}\,\boxed{3x^3} - 10x^2 - 7x + 30} \quad \overset{3x^2}{}$$

DEFINITION

Long division is a technique that achieves the division through a series of estimated quotients and repeated subtraction. The polynomial being divided is called the **dividend.** It is divided by the divisor to produce a quotient and possibly a remainder.

Multiply the entire divisor by the partial quotient you placed up top, and subtract.

$$\boxed{x} - 3\overline{\smash{\big)}\,\boxed{3x^3} - 10x^2 - 7x + 30} \quad \overset{3x^2}{}$$
$$\underline{-\left(3x^3 - 9x^2\right)}$$
$$-x^2$$

Bring down the next term and repeat the steps. $\frac{-x^2}{x} = -x$, so put that into the quotient, and multiply $x - 3$ by $-x$. Remember to subtract, and watch your signs.

$$\boxed{x} - 3\overline{\smash{\big)}\,3x^3 - 10x^2 - 7x + 30} \quad \overset{3x^2 - x}{}$$
$$\underline{3x^3 - 9x^2}$$
$$\boxed{-x^2} - 7x$$
$$\underline{-x^2 + 3x}$$
$$-10x$$

Once more and you should be done. Add −10 to the quotient, multiply, and subtract.

$$\require{enclose}\begin{array}{r}3x^2-x-10\\[-2pt]\boxed{x}-3\overline{\smash{\big)}\,3x^3-10x^2-7x+30}\end{array}$$

$$-\left(3x^3-9x^2\right)$$

$$-x^2-7x$$

$$-\left(-x^2+3x\right)$$

$$\boxed{-10x}+30$$

$$-\left(-10x+30\right)$$

$$0$$

Because you get a 0 remainder, you know $x-3$ is a factor and you know that $3x^3-10x^2-7x+30=(x-3)(3x^2-x-10)$. You can try to factor $3x^2-x-10$ or use the quadratic formula to find the other two zeros. In fact, $3x^2-x-10$ will factor, so here's the solution.

$$3x^3-10x^2-7x+30=0$$
$$(x-3)(3x^2-x-10)=0$$
$$(x-3)(3x+5)(x-2)=0$$
$$x-3=0 \quad 3x+5=0 \quad x-2=0$$
$$x=3 \qquad x=-\tfrac{5}{3} \qquad x=2$$

ALGEBRA TIP

Although it may not be necessary in every case, it's wise to insert zeros in the dividend for any missing powers. That will give you a spot for every possible power of the variable, allowing you to line up like terms.

The function has three real zeros: $x=3$, $x=2$, and $x=-\tfrac{5}{3}$.

The long division process may seem like a lot of work to test for a factor, and in fact there is an alternate method that you'll find easier. There are circumstances, however, when only long division will do the job, so it's important that you have both tools at your disposal. Long division is required when the divisor is second degree or higher, so if you need to divide $x^4-3x^3+6x^2-12x+8$ by x^2+4, for example, you'll need long division to find out that the quotient is x^2-3x+2.

$$
\require{enclose}
\begin{array}{r}
x^2 - 3x + 2 \\
x^2 + 4 \enclose{longdiv}{x^4 - 3x^3 + 6x^2 - 12x + 8}
\end{array}
$$

$$x^4 \qquad + 4x^2$$

$$- 3x^3 + 2x^2 - 12x$$

$$-3x^3 \qquad - 12x$$

$$2x^2 \qquad + 8$$

$$2x^2 \qquad + 8$$

$$0$$

CHECKPOINT

Divide using long division. Is the divisor a factor of the dividend?

16. $(x^3 - 3x^2 - 8x + 24) \div (x - 3)$ 19. $(2x^3 + 11x^2 + 12x - 9) \div (x^2 - 3)$

17. $(x^3 - 4x^2 - 3x + 12) \div (x + 2)$ 20. $(x^3 + 27) \div (x^2 - 3x + 9)$

18. $(2x^3 - 3x^2 - 8x + 12) \div (x - 6)$

Synthetic Division

If you need to test a first-degree factor, you can use long division, as you saw previously, or you can choose to use a technique known as *synthetic division*, or synthetic substitution (the latter being an alternate name that comes from the remainder theorem). You can use this technique to find the quotient and remainder when you divide by $x - c$, or—because the remainder is $f(c)$—you can use it to evaluate the function.

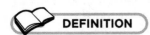

DEFINITION

Synthetic division, also known as synthetic substitution, is a technique for dividing a polynomial by a first-degree binomial, or for evaluating a polynomial for a particular value of the variable. It is based on the nested form of a polynomial: $ax^3 + bx^2 + cx + d = x(x(x(a) + b) + c) + d$.

To divide $x^3 + 8x^2 + 11x - 20$ by $x + 5$, for instance, start with just the coefficients of $x^3 + 8x^2 + 11x - 20$; the method trusts that you can remember the powers of the variable. If your polynomial has any missing terms, it's essential that you put in a 0 to hold the place. For the divisor, you'll actually use the corresponding zero. So if $x + 5 = 0$, $x = -5$. Place -5 in the box on the left, skip a line, draw a line, and bring the first coefficient down below the line.

$$\underline{-5|}1 \quad 8 \quad 11 \quad -20$$
$$\downarrow$$
$$\overline{1}$$

Multiply the number in the divisor box by the number below the line, and place the product under the next coefficient. Add, placing your sum below the line, and repeat. The process is repetitive: -5 times 1 plus 8 is 3; -5 times 3 is -15, plus 11, is -4; -5 times -4 plus -20 is 0. Box off that last result to mark it as the remainder.

$$\begin{array}{r|rrrr} -5 & 1 & 8 & 11 & -20 \\ & \downarrow & -5 & -15 & 20 \\ \hline & 1 & 3 & -4 & \boxed{0} \end{array}$$

The 0 remainder tells you that -5 is a zero, so $x + 5$ is a factor. The synthetic division also provides the coefficients of the quotient: 1, 3, and -4. You started with a third-degree polynomial and divided by a first degree, so the quotient will be a second degree—$1x^2 + 3x - 4$: $x^3 + 8x^2 + 11x - 20 = (x + 5)(x^2 + 3x - 4)$. From there, you can factor the quadratic, or use the quadratic formula.

$$x^3 + 8x^2 + 11x - 20 = (x + 5)(x^2 + 3x - 4) = (x + 5)(x + 4)(x - 1)$$

So the solutions are $x = -5$, $x = -4$, and $x = 1$.

Consider an example in which not all the solutions are real. Solving $x^4 - 3x^3 - 18x^2 + 90x - 100 = 0$ begins with a list of the possible rational zeros: $\pm 1, \pm 2, \pm 4, \pm 5, \pm 10, \pm 20, \pm 25, \pm 50, \pm 100$. You can expect four zeros, but not all necessarily real. Descartes' law of signs tells you the three sign changes indicate a maximum of three positive zeros, and the graph—if you have a graphing utility available—will narrow the list of possibilities. The best options look like $x = 2$ and $x = -5$. Use synthetic division to test them.

$$\begin{array}{r|rrrrr} 2 & 1 & -3 & -18 & 90 & -100 \\ & \downarrow & 2 & -2 & -40 & 100 \\ \hline & 1 & -1 & -20 & 50 & \boxed{0} \end{array}$$

The synthetic division tells you that $x = 2$ is a zero, and $x^4 - 3x^3 - 18x^2 + 90x - 100 = (x - 2)(x^3 - x^2 - 20x + 50)$. Test $x = -5$, using the reduced polynomial $x^3 - x^2 - 20x + 50$ and picking up just where you left off.

$$
\begin{array}{r|rrrrr}
2 & 1 & -3 & -18 & 90 & -100 \\
& \downarrow & 2 & -2 & -40 & 100 \\
\hline
-5 & 1 & -1 & -20 & 50 & \underline{|0} \\
& \downarrow & -5 & 30 & -50 & \\
\hline
& 1 & -6 & 10 & \underline{|0} &
\end{array}
$$

Now you have $x^4 - 3x^3 - 18x^2 + 90x - 100 = (x - 2)(x + 5)(x^2 - 6x + 10)$. Use the quadratic formula to solve $x^2 - 6x + 10 = 0$.

$$
x = \frac{6 \pm \sqrt{36 - 4(1)(10)}}{2(1)} = \frac{6 \pm \sqrt{-4}}{2} = \frac{6 \pm 2i}{2} = 3 \pm i
$$

The solutions of $x^4 - 3x^3 - 18x^2 + 90x - 100 = 0$ are $x = 2$, $x = -5$, $x = 3 + i$, and $x = 3 - i$.

CHECKPOINT

Use synthetic division to determine if the given value of x is a zero of the polynomial.

21. $f(x) = 2x^4 + 4x^3 - 2x^2 + 6$ given $x = 3$

22. $f(x) = 3x^3 - 8x^2 - 41x + 30$ given $x = -4$

23. $f(x) = 5x^4 - 3x^3 + 2x^2 - 3$ given $x = 2$

24. $f(x) = 3x^3 - 8x^2 + 5x + 16$ given $x = -1$

25. $f(x) = x^3 - 7x^2 + 2x + 40$ given $x = -2$

At times, when asked to find the zeros of a polynomial function, you are given one zero as a place to begin. If that zero is nonreal, you can choose to do synthetic division with the nonreal zero and then again with its conjugate. This process will be a bit messy, but if you work carefully and with patience, it will resolve itself. Here's how it might look if you were asked to find the zeros of $f(x) = x^4 - 3x^3 - 18x^2 + 90x - 100$, given that $3 + i$ is a zero.

$$\begin{array}{r|rrrrr} 3+i & 1 & -3 & -18 & 90 & -100 \\ & \downarrow & 3+i & -1+3i & -60-10i & 100 \\ \hline & 1 & i & -19+3i & 30-10i & \lfloor 0 \\ \end{array}$$

$$i(3+i)=-1+3i \quad (-19+3i)(3+i)=-60-10i \quad (30-10i)(3+i)=100$$

Then:

$$\begin{array}{r|rrrr} 3-i & 1 & i & -19+3i & 30-10i \\ & \downarrow & 3-i & 9-3i & -30+10i \\ \hline & 1 & 3 & -10 & \lfloor 0 \\ \end{array}$$

$$\downarrow \quad 3(3-i)=9-3i \quad -10(3-i)=-30+10i$$

This tells you $x^4 - 3x^3 - 18x^2 + 90x - 100 = (x - 3 - i)(x - 3 + i)(x^2 + 3x - 10)$. You can then choose the most efficient method to solve the quadratic.

Finding an equation with given zeros is simply a matter of writing the factors that correspond to the given zeros and multiplying. If you are asked for a third-degree polynomial with zeros that include -1 and $2i$, for instance, you need to realize that if $2i$ is a zero, $-2i$ is also a zero, so multiply $(x + 1)(x - 2i)(x + 2i)$ to get $(x + 1)(x^2 + 4) = x^3 + x^2 + 4x + 4$.

CHECKPOINT

Find all zeros of the polynomial function. One zero is given.

26. $f(x) = x^3 - 7x^2 + 7x + 15$ given $x = 5$

27. $g(x) = 2x^3 + 3x^2 - 18x + 8$ given $x = \frac{1}{2}$

28. $f(x) = 3x^4 + 7x^3 - 25x^2 - 63x - 18$ given $x = -2$

29. $g(x) = x^4 - 5x^3 + 10x^2 - 20x + 24$ given $x = 2i$

30. $f(x) = x^4 + 13x^2 + 36$ given $x = -3i$

ALGEBRA TIP

While synthetic division is an important tool, don't forget that you can switch to simply factoring—including factoring a cubic by grouping—whenever it seems possible.

Polynomial Inequalities

Polynomial inequalities, like $x^3 - 3x^2 \geq 4x - 12$, are best handled by moving all nonzero terms to one side and finding the real zeros of the function. Those zeros divide the real number line into sections, and the inequality's solution set is found by testing a point in each of those sections to see if the inequality is or is not true on that interval. The method takes its name, the *test point method*, from the fact that you try x-values from each section.

DEFINITION

The method of solving polynomial inequalities by finding the zeros and evaluating the function for values between zeros is known as the **test point method.**

The test point method was a critical skill before the advent of graphing calculators, and it remains useful, but the calculator gives you other options. A look at the graph of the polynomial on your calculator can take the place of the test point method if you just need a quick answer. Portions of the graph above the x-axis are greater than 0; portions below the x-axis are less than 0.

To solve $x^3 - 3x^2 \geq 4x - 12$, first rewrite it as $x^3 - 3x^2 - 4x + 12 = 0$, and find its factors. $x^3 - 3x^2 - 4x + 12 = (x - 3)(x^2 - 4) = (x - 3)(x + 2)(x - 2) \geq 0$, so the zeros are $x = 3$, $x = -2$, and $x = 2$. These zeros divide the real line into four intervals: $x < -2$, $-2 < x < 2$, $2 < x < 3$, and $x > 3$.

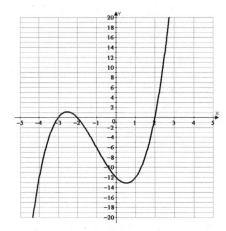

Choose a value from each interval and evaluate the function, or at least the sign of the function, at each value. A table like the following may be helpful. The sign of each factor for the chosen value of x is recorded. The sign of the function can be determined because it's the product of the factors.

Interval	Value	$(x + 2)$	$(x - 2)$	$(x - 3)$	$x^3 - 3x^2 - 4x + 12$
$x < -2$	-3	$-$	$-$	$-$	$-$
$-2 < x < 2$	0	$+$	$-$	$-$	$+$
$2 < x < 3$	2.5	$+$	$+$	$-$	$-$
$x > 3$	4	$+$	$+$	$+$	$+$

$x^3 - 3x^2 - 4x + 12 \geq 0$, or positive, in the interval $-2 < x < 2$ and on the interval $x > 3$.

 CHECKPOINT

Find the solution set for each polynomial inequality.

31. $x^3 - x^2 - 17x - 15 \geq 0$

32. $2x^3 + 3x^2 - 18x + 8 < 0$

33. $x^4 - 5x^2 + 6 \leq 0$

34. $4x^5 - 32x^4 + 48x^3 > 0$

35. $x^4 \leq 4x^2 + 45$

Graphs of Polynomial Functions

Unlike the functions you've graphed previously, polynomial functions do not have a single shape. As the degree of the polynomial changes, as the number and type of zeros change, the shape of the graph will change. There are some characteristics you can anticipate, however, and which you can use to help you sketch a graph.

Each real zero corresponds to an x-intercept and serves to anchor your graph. Nonreal zeros don't appear as x-intercepts, so you may have fewer x-intercepts than the degree of the function. You can also use the y-intercept, which is the constant term of the polynomial, to help you plan your sketch.

Linear functions, which are first-degree, never turn; quadratics, or second-degree functions, turn once. In general, you can anticipate that the number of *turning points* will be 1 less than the degree of the function. That will be clear when the zeros are real but harder to see when the function has complex zeros.

 **DEFINITION**

The **turning points** of the graph of a polynomial function are points at which the graph changes from increasing to decreasing or from decreasing to increasing.

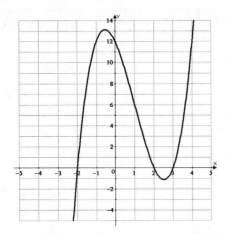
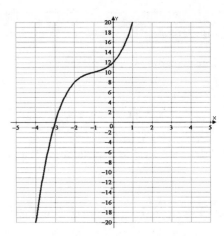

The graph on the left has three real zeros: –3, –2, and 2. It crosses the *x*-axis three times: at –3, at –2, and at 2. The graph on the right is also a cubic equation, so you would expect three solutions. It has one real zero, –3, which shows up as the single *x*-intercept at –3. The other two solutions are two imaginary numbers, 2*i* and –2*i*. You won't see them on the graph. The *x*-axis is a real number line, so nonreal zeros don't appear as intercepts. You'll need algebraic methods to find those two imaginary zeros.

To find the exact location of the turning points, a graphing utility is helpful. Manual calculation of those points requires skills from calculus.

The parent graph for the quadratic function and the cubic function can guide your sketch of the ends of the graph. Any polynomial function of even degree will rise on both ends if the lead coefficient is positive, and fall on both ends if it's negative, just like the parabola. All odd-degree polynomials will fall on the left and rise on the right if the lead coefficient is positive, like the cubic parent. If the lead coefficient is negative, the graph is reflected over the *x*-axis, so it rises on the left and falls on the right.

CALCULATOR CORNER

The graphing calculator can find the turning points of a polynomial function by using the maximum and minimum functions. Local maxima occur when the graph changes from increasing to decreasing, and local minima occur when it changes from decreasing to increasing. You'll need to provide the calculator with left and right bounds to define which turning point you're asking for.

CHECKPOINT

Sketch the graph of each function.

36. $y = 3x^3 - 7x^2 + 7x - 15$

37. $y = 2x^3 + 3x^2 - 18x + 8$

38. $y = x^4 - 5x^2 + 6$

39. $y = 4x^5 - 32x^4 + 48x^3$

40. $y = x^4 - 4x^2 - 45$

The Least You Need to Know

- Polynomial functions have as many zeros as their degree, if complex zeros are included, and their graphs have a number of turning points 1 less than the degree.

- Make a list of possible rational zeros from factors of the constant term divided by factors of the lead coefficient, use the number of sign changes to determine whether positive or negative real zeros are more probable, and start testing by synthetic division.

- If given a complex zero, you can do synthetic division with the zero and then with its complex conjugate, or you can multiply $(x - a - bi)(x - a + bi)$ to get a quadratic and use long division.

- Each time you find a factor by synthetic division or long division, use the other factor of lower degree than the original to search for other factors.

- Locate important points on the graph of a polynomial function by finding x- and y-intercepts, and use the number of turning points and the end behavior to shape the sketch.

Rational Functions

When you learned arithmetic, your first experiences focused on whole numbers in the decimal place value system. The concept of a fraction, and the adaptation of arithmetic to fractions, came later. It's not hard to draw a parallel between polynomials and those decimal system whole numbers. Just take a polynomial, like $4x^3 + 7x^2 + 8x + 3$, and replace x with 10, and you have $4(10)^3 + 7(10)^2 + 8(10) + 3 = 4(1,000) + 7(100) + 8(10) + 3 = 4,000 + 700 + 80 + 3$ or 4,783. It's natural then to ask what the algebra equivalent of a fraction might be, and how you would work with it. In this chapter, I take you through how to deal with rational expressions, as well as how to graph rational functions.

Rational Expressions and Their Arithmetic

The name *rational number*—or in algebra, *rational expression*—comes from the word *ratio*, which refers to a comparison by division. In the previous chapter, you looked at dividing polynomials by long and synthetic division, just as you once learned to divide whole numbers by long division and sometimes by "short division." In that original division and in your first polynomial division, you found a quotient and perhaps a remainder—an amount "left over." Those remainders can become the numerator of a fraction with the divisor as the denominator. For instance, you can say that 23 divided by 5 gives a quotient of 4 with a remainder of 3, or $23 \div 5 = 4\frac{3}{5}$.

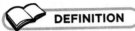

DEFINITION

A **rational expression** is the quotient of two polynomials, $\frac{p(x)}{q(x)}$, provided $q(x)$ is not a 0 polynomial.

A fraction is a rational number, a quotient of two integers. In algebra, a rational expression is the ratio or quotient of two polynomials. It can be as simple as $\frac{1}{x}$ or far more complicated, like $\frac{4x^2-9x+2}{x^3+5x^2-9x+1}$. The one restriction is that the polynomial in the denominator, no matter how complicated or how simple, must not be 0. This is a restriction that comes from the logic of arithmetic in how you define division. If you say that $\frac{a}{b} = c$, you're actually saying that $a = bc$. If you follow that reasoning, trying to divide by 0 means trying to find a number c that, when multiplied by 0, gives you a. That's impossible if a is any number other than 0. And it really doesn't work when a is 0 either, because then c could be any number. (If all that is starting to make your head hurt a bit, you understand why we never divide by 0.)

The practical concern when you work with rational expressions, however, is remembering and, in most cases, stating that there are values of the variable for which your rational expression may not be defined. The expression $\frac{1}{x}$ is only defined when x is not 0, and $\frac{2x-3}{x^2-4x+3}$ is undefined when $x^2 - 4x + 3 = 0$, so when $x = 1$ or when $x = 3$.

When you learned to work with fractions, you learned to simplify (or "reduce"—an unfortunate but popular word choice), add, subtract, multiply, and divide fractions, and you'll do all of those things with rational expressions. You'll also encounter what are called *complex fractions*—rational expressions in which the numerator or the denominator or both contain rational expressions or fractions. You never want to leave a fraction in a fraction, so there will be lots of cleaning up to do. Let's start with simplifying.

Simplifying Rational Expressions

When you look at the fraction $\frac{3}{6}$ and recognize it's equivalent to $\frac{1}{2}$, you probably think "3 and 6 can both be divided by 3." The theory behind that thought is that $\frac{3}{6} = \frac{3 \cdot 1}{3 \cdot 2}$, and that can be rewritten as $\frac{3}{3} \cdot \frac{1}{2} = 1 \cdot \frac{1}{2}$; you can often simplify a fraction without spelling out all those details. But if you're asked to simplify something like $\frac{x^2-4x-21}{x^2-12x+35}$, you might not be able to just look and know what to divide by, or what you get when you divide. So when simplifying a rational expression, you'll want to take it step by step as follows:

1. Factor the numerator, if possible: $\frac{x^2-4x-21}{x^2-12x+35} = \frac{(x-7)(x+3)}{x^2-12x+35}$

2. Factor the denominator, if possible: $\frac{x^2-4x-21}{x^2-12x+35} = \frac{(x-7)(x+3)}{(x-7)(x-5)}$

3. If the same factor appears in the numerator and denominator, cancel them because their quotient is 1: $\frac{x^2-4x-21}{x^2-12x+35} = \frac{\cancel{(x-7)}(x+3)}{\cancel{(x-7)}(x-5)} = \frac{x+3}{x-5}$

The domain of the expression in the example, the x-values for which it is defined, include all real numbers except those that make $x^2 - 12x + 35$ equal 0. When the denominator is in factored form, it's easier to see the domain is all reals except 5 and 7. However, be sure to look back to the original expression for the domain. Even though the factor of $x - 7$ canceled out, you couldn't have done the problem at all if the original expression was undefined.

ALGEBRA TRAP

When you start to cancel, it's easy to get on a "cross-out-everything-that-matches" run. Be careful. Canceling means removing a factor from the numerator and a matching factor from the denominator. You're removing a fraction equal to 1. Crossing out terms—expressions connected to others by addition or subtraction—doesn't accomplish the same thing. Never cancel terms, only factors.

CHECKPOINT

Give the domain of each rational expression and then simplify.

1. $\frac{7x}{21x^2}$

2. $\frac{5y+25}{y^2-25}$

3. $\frac{x^2-8x+15}{x^2-7x+10}$

4. $\frac{x^2-x-42}{x^2-49}$

5. $\frac{2y^2+7y+6}{2y^2+y-3}$

Adding and Subtracting

Adding fractions starts out easy. If the denominators are the same, add the numerators and keep the denominator: $\frac{2}{7} + \frac{3}{7} = \frac{5}{7}$. You should also simplify the result, if you can. The same is true with rational numbers. For example, $\frac{x+5}{x+1}$ and $\frac{x-3}{x+1}$ have the same denominator, so you just combine the numerators: $\frac{x+5}{x+1} + \frac{x-3}{x+1} = \frac{2x+2}{x+1}$. You then factor the numerator and cancel: $\frac{2x+2}{x+1} = \frac{2(x+1)}{x+1} = 2$. For this fraction, remember to state that the domain is all reals except -1. Even though it's simplified to 2, the original expression has a restricted domain.

Just as in arithmetic, the complications arise when your fractions have different denominators. Just as you change $\frac{1}{2}$ to $\frac{3}{6}$ and $\frac{1}{3}$ to $\frac{2}{6}$ so you can add $\frac{1}{2}+\frac{1}{3}=\frac{3}{6}+\frac{2}{6}=\frac{5}{6}$, you'll need to change the appearance but not the value of one or more of the rational expressions, so you have a common denominator. And just as simplifying was not as intuitive when rational expressions were involved, what the common denominator is and how to change the appearance of your rational expressions may not be obvious. The following is the plan of action to remember:

1. Factor each denominator, if possible: $\frac{x-3}{x^2+7x-44}+\frac{x+5}{x^2-16}=\frac{x-3}{(x+11)(x-4)}+\frac{x+5}{(x+4)(x-4)}$. Think about the domain before you move on. Both fractions must be defined, so you're operating on a domain of all reals except -11, -4, or 4.

2. Take note of any factors that appear in more than one denominator. You don't want to make the denominators any larger than necessary, so if there is a factor that already occurs in both denominators, don't repeat it: $\frac{x-3}{x^2+7x-44}+\frac{x+5}{x^2-16}=\frac{x-3}{(x+11)(\underline{x-4})}+\frac{x+5}{(x+4)(\underline{x-4})}$

3. Multiply each denominator by any factors that appear in other denominators but are not already present in this one: $\frac{x-3}{x^2+7x-44}+\frac{x+5}{x^2-16}=\frac{x-3}{(x+11)(x-4)\underline{(x+4)}}+\frac{x+5}{(x+4)(x-4)\underline{(x+11)}}$

4. Multiply each numerator by the same factors you introduced to the denominator:

$$\frac{x-3}{x^2+7x-44}+\frac{x+5}{x^2-16}=\frac{(x-3)(x+4)}{(x+11)(x-4)\underline{(x+4)}}+\frac{(x+5)(x+11)}{(x+4)(x-4)\underline{(x+11)}}$$

5. Expand (multiply out) the numerators, but let the denominators stay in factored form for now:

$$\frac{x-3}{x^2+7x-44}+\frac{x+5}{x^2-16}=\frac{x-3}{(x+11)(x-4)}\cdot\frac{x-4}{x-4}+\frac{x+5}{(x+4)(x-4)}\cdot\frac{x+11}{x+11}$$

$$=\frac{(x-3)(x-4)}{(x+11)(x-4)\underline{(x+4)}}+\frac{(x+5)(x+11)}{(x+4)(x-4)\underline{(x+11)}}$$

$$=\frac{x^2-7x+12}{(x+11)(x-4)(x+4)}+\frac{x^2+16x+55}{(x+4)(x-4)(x+11)}$$

6. Add the numerators by combining like terms:

$$\frac{x-3}{x^2+7x-44}+\frac{x+5}{x^2-16}=\frac{x^2+x-12}{(x+11)(x-4)\underline{(x+4)}}+\frac{x^2+16x+55}{(x+4)(x-4)\underline{(x+11)}}=\frac{2x^2+17x+43}{(x+11)(x-4)\underline{(x+4)}}$$

7. Check to see if the simplified numerator can be factored, and specifically, if it has any factors in common with the common denominator. Cancel, if possible. (In this example, it is not possible.)

8. If desired, multiply out the denominator:

$$\frac{x-3}{x^2+7x-44}+\frac{x+5}{x^2-16}=\frac{2x^2+17x+43}{(x+11)(x-4)(x+4)}=\frac{2x^2+17x+43}{(x+11)(x^2-16)}=\frac{2x^2+17x+43}{x^3+11x^2-16x-176}$$

Subtraction of rational expressions follows the same basic rules as addition, except, of course, that you subtract. There is one spot in the process at which particular care is necessary for subtraction, and that is the actual subtraction. You need to remember that the fraction bar, or vinculum, acts like a set of parentheses around the numerator, telling you to subtract (or change the sign and add) every term of the numerator that follows. So to subtract the fractions from the previous example, $\frac{x^2+x-12}{(x+11)(x-4)(x+4)} - \frac{x^2+16x+55}{(x+4)(x-4)(x+11)}$, you'd need to think of the problem as

$$\frac{x^2+x-12-\left(x^2+16x+55\right)}{(x+11)(x-4)(x+4)} = \frac{x^2+x-12-x^2-16x-55}{(x+4)(x-4)(x+11)} = \frac{-15x-67}{(x+4)(x-4)(x+11)}.$$

ALGEBRA TRAP

The biggest danger when subtracting rational expressions is forgetting that the fraction bar acts as a set of parentheses. The subtraction sign applies to every term of the numerator that follows it, so be careful to subtract the entire numerator. It may help to distribute the minus sign and then add.

CHECKPOINT

Perform addition or subtraction as indicated and leave the answers in simplest form.

6. $\frac{3x+2}{x} + \frac{5x+7}{2x}$

7. $\frac{a}{a+3} + \frac{2a}{3a+9}$

8. $\frac{2}{x} + \frac{3}{x+1}$

9. $\frac{x+2}{x+3} - \frac{x+3}{x+4}$

10. $\frac{y-2}{y^2-5y+4} - \frac{y-2}{y^2-16}$

Multiplying Rational Expressions

The fundamental rule for multiplying fractions—numerator times numerator, denominator times denominator—also applies to rational expressions. And if you're multiplying $\frac{3x}{x+2} \cdot \frac{x-1}{x-2}$, for instance, it's a workable plan: $\frac{3x}{x+2} \cdot \frac{x-1}{x-2} = \frac{3x^2-3x}{x^2-4}$ with a domain of all reals except 2 and –2.

On the other hand, if you need to multiply $\frac{2x^2+5x-3}{x^2+5x+6} \cdot \frac{x^2+4x+4}{4x^2+8x-5}$, the multiplication is a significant undertaking, and then you'll have to think about simplifying. You'll probably be looking for a better way. The better plan you're seeking amounts to doing the simplifying first. Follow the three most important rules for multiplying rational expressions: 1) factor, 2) factor, and 3) factor.

Don't even think of multiplying right away; instead, factor. Factor the numerators, factor the denominators, and factor anything that can be factored. In the case of the example, $\frac{2x^2+5x-3}{x^2+5x+6} \cdot \frac{x^2+4x+4}{4x^2+8x-5} = \frac{(2x-1)(x+3)}{(x+3)(x+2)} \cdot \frac{(x+2)(x+2)}{(2x-1)(2x+5)}$. This is the moment to consider the domain. For this expression, you're working on all reals except –3, –2.5, –2, and 0.5.

Multiplying the numerators and multiplying the denominators will give you a single rational expression equal to $\frac{(2x-1)(x+3)(x+2)(x+2)}{(x+3)(x+2)(2x-1)(2x+5)}$, but you may notice there are factors in the numerator that match factors in the denominator. Rearranging the factors of the numerator a bit lets you cancel: $\frac{\cancel{(x+3)}\,\cancel{(x+2)}\,\cancel{(2x-1)}(x+2)}{\cancel{(x+3)}\,\cancel{(x+2)}\,\cancel{(2x-1)}(2x+5)}$. When you have canceled them, your product has simplified down to $\frac{x+2}{2x+5}$, and you didn't need to do any multiplying at all. It won't always be that easy, but factoring first and canceling is usually a better strategy than multiplying right away.

CHECKPOINT

Multiply and leave the answers in simplest form.

11. $\frac{x^2-1}{6-3x} \cdot \frac{12-6x}{2x+2}$

12. $\frac{x^2+7x+10}{x^2-2x+1} \cdot \frac{x^2-6x+5}{x^2-25}$

13. $\frac{x^2-16x+28}{2x-28} \cdot \frac{3x^3}{6x^2-12x}$

14. $\frac{x^2-5x+6}{x^2-7x+10} \cdot \frac{x^2-4x-5}{x^2-2x-3}$

15. $\frac{x^2+x-42}{x^2-36} \cdot \frac{x^2-49}{x^2-x-42}$

Dividing by a Rational Expression

The rule for dividing by a rational expression is simple: don't. To divide by a rational expression, multiply by the *multiplicative inverse* or reciprocal of the expression. Two expressions are multiplicative inverses, or reciprocals, if they multiply to 1. For a rational expression like $\frac{x+3}{x-7}$, for example, that's just $\frac{x-7}{x+3}$. Inverting the divisor is simple enough, but it does introduce a new restriction on the domain. If you were dividing $\frac{x^2-9}{x^2-8x+7} \div \frac{x+3}{x-7}$, the initial problem is only defined for real numbers other than 7 and 1. When you rewrite the problem as $\frac{x^2-9}{x^2-8x+7} \cdot \frac{x-7}{x+3}$, you also have to consider that $x \neq -3$. That gives you three values not in the domain: −3, 1, and 7.

DEFINITION

The **multiplicative inverse,** or reciprocal, inverts, or flips, the fraction or expression, exchanging the numerator and denominator. Two rational numbers are multiplicative inverses if their product is 1. Two rational expressions are multiplicative inverses if they multiply to 1 over a domain on which both expressions are defined.

Once you have inverted the divisor, you multiply. You do that by factoring numerators and denominators, and canceling where you can: $\frac{x^2-9}{x^2-8x+7} \cdot \frac{x-7}{x+3} = \frac{\cancel{(x+3)}(x-3)}{(x-1)\cancel{(x-7)}} \cdot \frac{\cancel{x-7}}{\cancel{x+3}}$. Make a habit of inverting the divisor before you begin to factor. Once you start factoring, it's easy to get involved in that process and begin canceling, and forget to invert. Finally, you multiply, if necessary, remaining factors in the numerators, and then multiply factors in the denominators:

$$\frac{x^2-9}{x^2-8x+7} \cdot \frac{x-7}{x+3} = \frac{\cancel{(x+3)}(x-3)}{(x-1)\cancel{(x-7)}} \cdot \frac{\cancel{x-7}}{\cancel{x+3}} = \frac{x-3}{x-1}.$$

CHECKPOINT

Divide and leave the answers in simplest form.

16. $\frac{5x^2-20}{6} \div \frac{x^2-4x+4}{3}$

17. $\frac{5x^2-20}{x^2-5x+6} \div \frac{6x+12}{3x-6}$

18. $\frac{2x^2-x}{4x^2-1} \div \frac{4x}{8x+4}$

19. $\frac{1-16y^2}{y^4-16} \div \frac{y-4y^2}{y^2-4y+4}$

20. $\frac{a^2+4a-45}{4a-20} \div \frac{a^2-81}{3a-3}$

Complex Fractions

If you were doing arithmetic, you wouldn't leave an answer as $\frac{2/3}{4}$. That construction of a fraction within a fraction is called a *complex fraction,* and is not very satisfying. To simplify $\frac{2/3}{4}$, you need to do what it says: $\frac{2}{3} \div 4 = \frac{2}{3} \div \frac{4}{1} = \frac{2}{3} \cdot \frac{1}{4} = \frac{1}{6}$.

DEFINITION

A **complex fraction** is a fraction that contains one or more fractions in its numerator and/or denominator.

Faced with the rational expression version of a complex fraction, you can use the same strategy: do the arithmetic it talks about. For example, $\frac{\frac{2}{x-3}+5}{\frac{4}{x-3}-1}$ is asking you to do quite a bit of arithmetic. Add 5 to the rational expression $\frac{2}{x-3}$, subtract 1 from $\frac{4}{x-3}$, and then divide your results. Taken a bit at a time, it's doable.

First: $\frac{2}{x-3}+5 = \frac{2}{x-3} + \frac{5x-15}{x-3} = \frac{5x-13}{x-3}$

Next: $\frac{4}{x-3}-1 = \frac{4}{x-3} - \frac{x-3}{x-3} = \frac{4-x+3}{x-3} = \frac{7-x}{x-3}$

Finally: $\frac{5x-13}{x-3} \div \frac{7-x}{x-3} = \frac{5x-13}{x-3} \cdot \frac{x-3}{7-x} = \frac{5x-13}{7-x}$

You can, step by step, arrive at the fact that $\frac{\frac{2}{x-3}+5}{\frac{4}{x-3}-1} = \frac{5x-13}{7-x}$.

There is another tactic you can use that you may find more convenient. If you look over the complex fraction and find a common denominator for all the little interior fractions, you can arrive at the simplified version faster. The complex fraction in the example has two interior fractions that both have denominators of $x - 3$. Take the complex fraction and multiply the numerator and the denominator by $x - 3$. That means you're multiplying by 1, so the appearance changes but not the value.

$$\frac{\frac{2}{x-3}+5}{\frac{4}{x-3}-1} \cdot \frac{x-3}{x-3} = \frac{(x-3)\left(\frac{2}{x-3}\right)+5(x-3)}{(x-3)\left(\frac{4}{x-3}\right)-1(x-3)} = \frac{2+5x-15}{4-x+3} = \frac{5x-13}{7-x}$$

If the denominators are not all the same, it may take a little more work. To simplify $\frac{\frac{1}{x-1}+2}{2-\frac{1}{x+1}}$, for instance, you need to use the common denominator $(x - 1)(x + 1)$, and carefully multiply the numerator and denominator by that expression.

$$\frac{\frac{1}{x-1}+2}{2-\frac{1}{x+1}} \cdot \frac{(x-1)(x+1)}{(x-1)(x+1)} = \frac{(x-1)(x+1)\left(\frac{1}{x-1}\right)+2(x-1)(x+1)}{2(x-1)(x+1)-(x-1)(x+1)\left(\frac{1}{x+1}\right)} = \frac{(x-1)\cdot1+2(x-1)(x+1)}{2(x-1)(x+1)-(x-1)\cdot1}$$

Make sure you see the denominators cancel before you rush to expand the products.

$$\frac{(x-1)\cdot1+2(x-1)(x+1)}{2(x-1)(x+1)-(x-1)\cdot1} = \frac{x-1+2x^2-2}{2x^2-2-x+1} = \frac{2x^2+x-3}{2x^2-x-1}$$

Finally, check to see if this expression can be simplified further.

$$\frac{2x^2+x-3}{2x^2-x-1} = \frac{(2x+3)(x-1)}{(2x+1)(x-1)} = \frac{2x+3}{2x+1}$$

ALGEBRA TIP

Before you multiply out the numerator, look over the numerator and denominator to see if it's possible to factor them at that point. An expression like $\frac{(x-1)\cdot1+2(x-1)(x+1)}{2(x-1)(x+1)-(x-1)\cdot1}$ requires a lot of simplifying and will then have to be factored again.

Instead, notice a common factor in all the terms: $\frac{\left[(x-1)\cdot1\right]+\left[2(x-1)(x+1)\right]}{\left[2(x-1)(x+1)\right]-\left[(x-1)\cdot1\right]}$.

If you factor that out, and cancel it out, simplifying will be easier:
$\frac{(x-1)\left[1+2(x+1)\right]}{(x-1)\left[2(x+1)-1\right]} = \frac{1+2(x+1)}{2(x+1)-1} = \frac{2x+3}{2x+1}$.

CHECKPOINT

Simplify each complex fraction as completely as possible.

21. $\dfrac{4+\frac{2}{x}}{\frac{4}{x}-2}$

22. $\dfrac{\frac{2}{3x}-\frac{3}{2x}}{\frac{5}{2x}+\frac{2}{3x}}$

23. $\dfrac{\frac{1}{x}-1}{1+\frac{1}{x}}$

24. $\dfrac{\frac{3}{x-1}+5}{\frac{1}{x-5}+3}$

25. $\dfrac{\frac{1}{x-3}+\frac{x}{x+2}}{\frac{x}{x-3}-\frac{1}{x+2}}$

Solving Rational Equations

A *rational equation* is simply an equation that contains one or more rational expressions. Solving a rational equation combines the skills you use to operate with rational expressions and your techniques for solving linear and quadratic equations. You'll also have to consider the restrictions on the domain of the rational expressions when you look at your results. Checking for extraneous solutions will be essential. You should make a note of excluded values before you begin, but as you get involved in the work of solving, it's easy to forget that some values are excluded from the domain. Always take a minute to check back with the original equation to make sure your solutions are values in the domain.

DEFINITION

A **rational equation** is an equation that contains one or more rational expressions.

There are two common techniques for solving rational equations: cross-multiplying and multiplying through.

Cross-Multiplying

In arithmetic, when you learned about ratios and proportions, you learned to solve a proportion—an equation that said two ratios are equal—by a technique called *cross-multiplying*. While this technique has limited usefulness, it is often the simplest, fastest method when it can be used. For example, if you were looking for the missing term in the proportion $\frac{5}{7} = \frac{154}{x}$, you were able to apply what is officially called the *means-extremes property*. It says that in any proportion, the product of the means (the first and last terms) is equal to the product of the extremes (the two middle terms). In this example, that means that if $\frac{5}{7} = \frac{154}{x}$, then $5x = 7(154)$. You could achieve the same result by starting with $\frac{5}{7} = \frac{154}{x}$, multiplying both sides by x, and multiplying both sides by 7, but generally you just think of it as cross-multiplying.

You can solve a rational equation by cross-multiplying if the equation is two equal rational expressions, or you can conveniently simplify the equation to two equal rational expressions. Cross-multiplying may result in a linear equation, a quadratic equation, or a polynomial of higher degree.

To solve $\frac{x+5}{x-3} = \frac{x+1}{x-2}$, first take a moment to think about the domain. You need to eliminate 2 and 3 from the domain of this equation, because either of those would make a denominator 0. The equation has only one expression on each side, so you can cross-multiply: $\frac{x+5}{x-3} = \frac{x+1}{x-2}$ becomes $(x + 5)(x - 2) = (x - 3)(x + 1)$. Expand each side, collect like terms, and solve.

$$(x+5)(x-2)=(x-3)(x+1)$$
$$x^2 +3x-10 = x^2 -2x-3$$
$$5x = 7$$
$$x = \tfrac{7}{5}$$

Before you decide you're done, remember to check your solution against the domain of the equation. Your domain was all real numbers except 2 or 3, so $\frac{7}{5}$ is fine.

ALGEBRA TIP

Resist the temptation to simplify expressions while adding or subtracting. You just changed the fractions to a common denominator. If you simplify now, you'll go around in a circle and find yourself back where you started.

If your equation is more complicated, you could theoretically simplify each side down to a single fraction, and then cross-multiply. In many cases, however, this is a tedious task and may produce a third- or fourth-degree polynomial equation to solve. It's also likely to result in extraneous solutions. So let's look at another method that's more efficient.

CHECKPOINT

Solve by cross-multiplying.

26. $\frac{4}{t-6} = \frac{3}{t}$ 27. $\frac{x-3}{x+7} = \frac{x-2}{x+4}$ 28. $\frac{3}{x-4} = \frac{x+4}{3}$

29. $\frac{4}{x+1} = \frac{x}{5}$ 30. $\frac{3x+1}{x-1} = \frac{x+2}{x-2}$

Multiplying Through

The second method of solving rational equations is the more widely useful. Generally referred to as *multiplying through*, it is a method that eliminates all denominators by multiplying each term on both sides by a common denominator of all the rational expressions in the equation.

Consider solving the equation $\frac{2}{x-3} + \frac{x}{x+5} = \frac{2x-1}{3x-9}$. Take a minute to factor $3x - 9$, and you'll be able to see two important things.

$$\frac{2}{x-3} + \frac{x}{x+5} = \frac{2x-1}{3(x-3)}$$

 **CALCULATOR CORNER**

When entering rational functions on your calculator, enclose the numerator and the denominator each in their own set of parentheses to avoid any order of operations confusion.

The domain of this equation is all real numbers except 3 and –5, and the common denominator for the three rational expressions is $3(x - 3)(x + 5)$. Multiply each side of the equation by this common denominator, distributing on the left side. You'll see that for each fraction, the denominator will cancel with factors in the common denominator.

$$3(x-3)(x+5)\left[\tfrac{2}{x-3} + \tfrac{x}{x+5}\right] = 3(x-3)(x+5)\left[\tfrac{2x-1}{3(x-3)}\right]$$

$$3(x-3)(x+5)\left[\tfrac{2}{x-3}\right] + 3(x-3)(x+5)\left[\tfrac{x}{x+5}\right] = 3(x-3)(x+5)\left[\tfrac{2x-1}{3(x-3)}\right]$$

$$3(x+5)\cdot 2 + 3(x-3)\cdot x = (x+5)(2x-1)$$

Do the remaining multiplication and simplify each side.

$$6x + 30 + 3x^2 - 9x = 2x^2 + 9x - 5$$
$$3x^2 - 3x + 30 = 2x^2 + 9x - 5$$

You're left with a quadratic equation, so gather all terms on one side equal to 0, and factor (or use the quadratic formula).

$$x^2 - 12x + 35 = 0$$
$$(x-5)(x-7) = 0$$
$$x-5 = 0 \qquad x-7 = 0$$
$$x = 5 \qquad\quad x = 7$$

Check these potential solutions against the restrictions on the domain. The equation is defined for all reals except 3 and −5, so both of these solutions are in the domain.

CHECKPOINT

Solve each equation. Be sure to check for extraneous solutions.

31. $\frac{2x-5}{6} - \frac{1}{4} = \frac{3-x}{12}$

32. $\frac{3}{a} - \frac{1}{2a} + \frac{1}{3a} = 6$

33. $\frac{1}{x} + \frac{1}{x-1} = \frac{5}{6}$

34. $\frac{7}{x+2} = 2 - \frac{3}{x-2}$

35. $\frac{6}{x-1} + \frac{8}{x} = \frac{5}{x^2+x}$

How to Graph Rational Functions

Rational functions are probably the most complex graphs you will sketch, because of the denominators that cause restrictions to the domain. Because there are *discontinuities,* or values for which the function is undefined, there will be breaks in the graph. Like the hyperbolas you saw in Chapter 9, rational functions generally are made up of two or more sections that make up the graph of a single function.

DEFINITION

A **discontinuity** is a break in the graph of a function that occurs at a value excluded from the domain.

Domain

Before you begin to sketch, you need to think about the domain of the function, or actually, about what values are not in the domain. While there are rational functions that are defined for all real numbers, they are not the typical rational functions. The parent function for the rational family is the graph of $y = \frac{1}{x}$, which has a domain of all reals except 0. This break in the domain means a break in the graph. There is no point on the graph when $x = 0$. There cannot be, because there is no number equal to $\frac{1}{0}$. The graph is therefore in two pieces, as you see in the following.

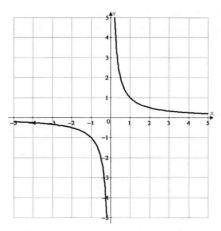

Rational functions of the form $y = \frac{ax+b}{cx+d}$ have only one discontinuity, and will be transformations of this parent graph. If there is more than one factor in the denominator, resulting in more than one discontinuity, the graph will have more than two pieces. The leftmost and rightmost pieces will resemble the two sections of the parent graph, but other pieces will be harder to predict. If the numerator is a higher degree than the denominator, the ends of the graph will rise and fall, rather than flattening out. And if the numerator and denominator have a factor in common, that factor will result in a discontinuity that is just a tiny hole in the graph.

Asymptotes

The parent graph exhibits a common characteristic of rational functions: asymptotic behavior. Close to $x = 0$, the function doesn't simply stop and jump over the discontinuity. On the positive side, it becomes very large, and on the negative side, very small. The change is so steep that the graph becomes almost (but never actually) vertical, and it continues without ever stopping. We say that one piece of the graph, on the positive side of 0, approaches infinity as x approaches 0 from the right, and the other piece, on the negative side, approaches negative infinity as x approaches 0 from the left. In symbols, this is shown as $f(x) \to \infty$ as $x \to 0^+$ and $f(x) \to -\infty$ as $x \to 0^-$.

In this case, the vertical line $x = 0$—the y-axis—is called a *vertical asymptote* of the graph. A vertical asymptote is a line the graph approaches closer and closer, but never touches or crosses. It has the equation $x =$ the x-value that is not in the domain. The graph cannot touch or cross the asymptote because the function is not defined at that x-value, but it increases to infinity or decreases to negative infinity as it nears the asymptote.

The parent graph also has a *horizontal asymptote*. The graph of $y = \frac{1}{x}$ gets extremely close to the line $y = 0$ as x gets very large or very small. The function never equals 0; the graph does not touch the line $y = 0$, but it gets closer and closer as it moves away from the origin. The end behavior of this graph is to approach the horizontal asymptote $y = 0$. End behavior is something of a misnomer, because the graph doesn't end, but for very large or very small values of x, the value of the function approaches 0 from above on one side and from below on the other.

DEFINITION

> A **vertical asymptote** is a vertical line at the point of discontinuity. The portions of the graph above and below the vertical asymptote either increase or decrease without bound as they get close to the vertical asymptote. A **horizontal asymptote** is a horizontal line that the ends of the graph of a rational function will approach very closely, either from above or below.

If the degree of the numerator is smaller than the degree of the denominator, the horizontal asymptote will be $y = 0$—the x-axis. If the numerator and denominator are the same degree, the horizontal asymptote is $y =$ a nonzero constant, which you can find by dividing the lead coefficient of the numerator by the lead coefficient of the denominator. For instance, the horizontal asymptote for $y = \frac{5x-3}{2x+1}$ is $y = \frac{5}{2}$.

When the degree of the numerator is the larger one, the function does not have a horizontal asymptote, but it may have a slant, or oblique, asymptote. To start, divide the numerator by the denominator; you'll probably want to use long division. If the quotient is a first-degree expression, $y =$ that expression is the slant asymptote. For example, to find the equation of the oblique asymptote of the rational function $f(x) = \frac{x^3-3x^2-4}{x^2-2x-24}$, do the long division.

$$
\begin{array}{r}
x - 1 \\
x^2 - 2x - 24 \overline{\smash{)}\,x^3 - 3x^2 + 0x - 4} \\
\underline{x^3 - 2x^2 - 24x} \\
-x^2 + 24x - 4 \\
\underline{-x^2 + 2x + 24} \\
22x - 28
\end{array}
$$

The slant asymptote is $y = x - 1$.

When you are given a rational function, have a plan. Let's look at $y = \frac{2x^2-7x+3}{x^2+2x-15}$ to see how that's accomplished.

First, always look at the denominator to find the domain. Factoring the denominator, if possible, will help you determine what values must be excluded. Do that before you do any simplifying.

$$y = \frac{2x^2-7x+3}{x^2+2x-15} = \frac{2x^2-7x+3}{(x+5)(x-3)}$$

So the domain is all real numbers except 3 and –5.

Once you've established the domain, check to see if the numerator and denominator have any factors in common, and if so, cancel. If you are able to cancel, note that the domain does not change. No excluded value ever returns to the domain, but if a factor cancels, the excluded value it created will cause just a tiny discontinuity, commonly called a *hole*, rather than a vertical asymptote.

$$y = \frac{2x^2-7x+3}{x^2+2x-15} = \frac{2x^2-7x+3}{(x+5)(x-3)} = \frac{(2x-1)(x-3)}{(x+5)(x-3)} = \frac{2x-1}{x+5}$$

The discontinuity at $x = 3$ will be just a tiny hole, but otherwise the function $y = \frac{2x^2-7x+3}{x^2+2x-15}$ acts like $y = \frac{2x-1}{x+5}$.

Now identify the vertical asymptote(s). Remember to specify each vertical asymptote by an equation. The vertical asymptote will have an equation of the form $x =$ an excluded value. $y = \frac{2x^2-7x+3}{x^2+2x-15}$ will have a vertical asymptote $x = -5$.

Finally, compare the degree of the numerator to the degree of the denominator. If the degree of the numerator is lower, the function has a horizontal asymptote of $y = 0$. If the degrees of the numerator and denominator are the same, divide the coefficients of the lead terms. The horizontal asymptote will be $y =$ the quotient of the lead coefficients. If the degree of the numerator is higher than the degree of the denominator, divide the numerator by the denominator, using long division if necessary, and ignore the remainder. If the quotient is a first degree expression, the slant asymptote is $y =$ that quotient. If the quotient has a higher degree, the function does not have a horizontal asymptote or a slant asymptote. The numerator and denominator of $y = \frac{2x^2-7x+3}{x^2+2x-15}$ are both second degree, so the horizontal asymptote is $y = \frac{2x^2}{x^2} = \frac{2}{1}$, or simply $y = 2$.

Here's another example. The rational function $y = \frac{4x-1}{x-1}$ has a domain of all real numbers except 1. Nothing factors, so nothing cancels, and the vertical asymptote is $x = 1$. The degrees of the numerator and the denominator are the same, so the horizontal asymptote is $y = \frac{4x}{x} = \frac{4}{1}$ or $y = 4$. Although you can see there is no slant asymptote, take a moment to divide the numerator by the denominator anyway. You get $y = 4 + \frac{3}{x-1}$; that is a quotient of 4 and a remainder of 3. Notice that $y = 4$ is the horizontal asymptote. You can divide the numerator by the denominator to find a horizontal asymptote as you do for a slant asymptote, and if you happen to be given the function in this form, you can spot the horizontal asymptote as $y =$ the constant.

CHECKPOINT

Give the domain of each function. Find all the asymptotes: vertical, horizontal, and oblique (if they exist).

36. $f(x) = \frac{2}{x-3} + 1$

37. $g(x) = \frac{2x-3}{x+1}$

38. $f(x) = \frac{x}{x^2-4}$

39. $g(x) = \frac{x^2-1}{x+4}$

40. $f(x) = \frac{2x-5}{x+2}$

Intercepts

The parent graph has no intercepts because both axes are its asymptotes, and it doesn't touch them. When the parent graph is translated, the new functions will have intercepts and you can use them to help you locate the graph. Let's look at an example.

To graph $f(x) = \frac{x+3}{x-1}$, note that the domain is all reals except $x = 1$, so there is a vertical asymptote of $x = 1$. The numerator and denominator have the same degree, so the horizontal asymptote is $y = \frac{1}{1} = 1$. The x-intercept is the solution of $0 = \frac{x+3}{x-1}$, or $(-3, 0)$; the y-intercept can be found by substituting 0 for x, $y = \frac{0+3}{0-1} = -3$, so $(0, -3)$. Sketching the asymptotes and plotting the intercepts will help you locate the pieces of the graph.

To better see the transformations, divide the numerator by the denominator, expressing it as quotient + $\frac{\text{remainder}}{\text{divisor}}$. $f(x) = \frac{x+3}{x-1} = 1 + \frac{4}{x-1}$ tells you that the parent graph was shifted up 1, stretched by a factor of 4, and shifted right 1 unit.

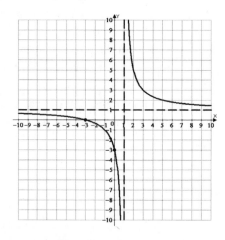

The graph of $f(x) = \frac{2x^2-5x-3}{x^2-4}$, or $f(x) = \frac{(2x+1)(x-3)}{(x+2)(x-2)}$, has two vertical asymptotes, $x = 2$ and $x = -2$, and a horizontal asymptote of $y = 2$. It has two x-intercepts, $\left(-\frac{1}{2},0\right)$ and $(3, 0)$, and a y-intercept $\left(0,\frac{3}{4}\right)$. Plotting those will give you a start on the graph, but you'll probably need to find a few more points before you're confident of the shape of the graph.

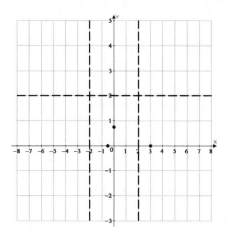

ALGEBRA TIP

The graph of a rational function never crosses a vertical asymptote but can sometimes cross a horizontal asymptote. The horizontal asymptote tells you about the end behavior of the graph—about what happens as x becomes very large or very small—but for smaller values, the graph may cross the horizontal asymptote.

The graph cannot cross the vertical asymptotes, so it will have three pieces. The rightmost piece will pass through the x-intercept at 3, and approach the vertical asymptote and the horizontal asymptote. You can see where that will go.

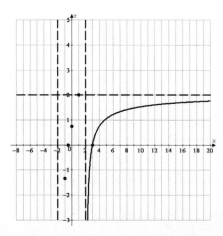

There will be a similarly shaped piece on the left, but it may be above or below the horizontal asymptote, so test a point to locate it. If $x = -3$, $f(x) = \frac{2(9)-5(-3)-3}{(9)-4} = \frac{18+15-3}{5} = 6$, which tells you the leftmost piece is above the horizontal asymptote.

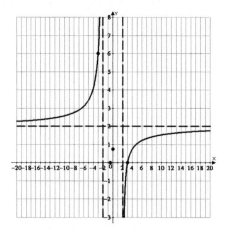

The middle piece, between the two vertical asymptotes, is the least predictable, so don't be afraid to test points. It will approach both vertical asymptotes, but can cross the horizontal asymptote.

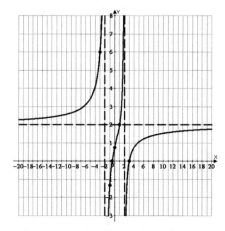

CHECKPOINT

Sketch a graph of each function. Show asymptotes and intercepts.

41. $f(x) = \frac{2}{x-3} + 1$

42. $g(x) = \frac{2x-3}{x+1}$

43. $f(x) = \frac{2x-5}{x+2}$

44. $g(x) = \frac{x^2-1}{x+4}$

45. $f(x) = \frac{x}{x^2-4}$

The Least You Need to Know

- A rational expression is the quotient of two polynomials, a rational equation is an equation containing one or more rational expressions, and a rational function is a function that contains a rational expression.

- When you simplify, multiply, or divide rational expressions, factor all polynomials that can be factored. Invert any divisors, and then cancel matching factors in the numerator and denominator.

- To add or subtract rational expressions, factor the denominators, determine the lowest common denominator, and multiply the numerator and denominator of each expression by any factors of the lowest common denominator missing from the denominator. Finally, multiply out the numerators and collect like terms.

- Solve rational equations containing two equal fractions by cross-multiplying. Solve more complicated rational equations by multiplying through by a common denominator to clear all the fractions.

- Any value that makes a denominator 0 must be eliminated from the domain. At each of those discontinuities, there will be a vertical asymptote, unless there is a matching factor in the numerator to cancel the denominator factor, causing the break. In that case, there will only be a tiny hole.

- If the degree of the numerator is less than the degree of the denominator, the horizontal asymptote is $y = 0$. If the degrees are the same, the horizontal asymptote is y equal to the ratio of the lead coefficients. If the degree of the numerator is larger than the degree of the denominator, find the slant asymptote by long division.

Calculating and Counting

Remember exponents? You learned them first in arithmetic, and then they started showing up in quadratics and higher-degree polynomials. Along the way, you worked with roots, which seemed to be the answer to undoing those exponents. Now you'll see variables moving into the exponent position, which opens new territory to explore and eliminates roots as a workable inverse.

So in this part, you look at exponents and logarithms, an idea that started as a calculating device. By the time calculators and computers came along, however, they had proved useful in other ways. So they're still around as the inverse of exponentiation, and therefore helpful when you want to explore exponential growth and decay.

You also look at counting in this part, and at the probability of different occurrences. And in the end, you find a bonus to bring the circle around to polynomials again.

Exponential and Logarithmic Functions

Exponents seem to pop up everywhere in algebra, primarily in quadratics and polynomials. Around the same time you learn about exponents, you're introduced to roots as a way to undo the work of an exponent. That works fine, as long as your variable is the base of the power and the exponent is a number. But what if the exponent is a variable? If $x^3 = 8$, you can take the cube root of both sides to find that $x = 2$. But if $3^x = 8$, you can't take the xth root of 8, because you don't know what x is.

In this chapter, you investigate exponential functions—functions that have the variable in the exponent position. You'll also explore their inverses, logarithmic functions.

Logarithms

Logarithms—or logs, as they're commonly called—take their name from John Napier's work on "logical arithmetic," an idea based on the fact that multiplying and dividing powers is as simple as adding or subtracting exponents. Designed to simplify calculations, logarithms have given way to calculators and computers for that sort of work; however, they continue to have other important applications.

In This Chapter

- Understanding the relationship of exponents and logarithms
- Using properties of logarithms to expand or condense log expressions
- Solving exponential and logarithmic equations and common problems
- Graphing exponential and logarithmic functions

The official definition of a logarithm connects it back to an exponential expression. If b and n are positive real numbers and $n = b^x$, then $\log_b n = x$. You would read that as "the log, base b, of n is x," but the better way to think of it might be "the exponent I would put on b to make n is x." For instance, if $5^3 = 125$, the exponent you would put on 5 to make 125 is 3, or $\log_5 125 = 3$. To make $\frac{1}{8}$, you could start with a base of 2 and put on an exponent of -3, so $\log_2 \frac{1}{8} = -3$.

Extending that idea a bit, if $y = b^x$, then x is the exponent you put on b to make y, so $\log_b y = x$. Going back to an earlier example, if $3^x = 8$, x is the exponent you put on 3 to make 8, or $\log_3 8 = x$. Now you have a way to solve that equation and others like it.

Of course, you need to find out the number $\log_3 8$ represents. Originally, that required going to a book of tables to look things up, but with today's calculators, it's a simpler task. Your calculator has a key marked log, which is $\log_{10}(\ )$. The number whose log you want goes in the parentheses. You'll also see a key marked ln, which is the natural log, or $\log_e(\ )$. The number $e \approx 2.71828\ldots$. That may seem a surprising number to use as the base of a log, but that will be explained later. The immediate issue is that you don't have a $\log_3(\ )$ key.

It's not practical to have a different key for every base, just as it wasn't practical to have a table for base. Instead, you'll use the *change-of-base rule:* $\log_3 8 = \frac{\log_{10} 8}{\log_{10} 3} = \frac{\ln 8}{\ln 3}$. The rule basically tells you to take the log of the number and divide by the log of the base you want. You can use your log key or your ln key, as long as you use the same key for the numerator and the denominator: $\log_3 8 = \frac{\log_{10} 8}{\log_{10} 3} \approx 1.89$.

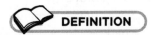 **DEFINITION**

A **logarithm** is the exponent that should be placed on a specified base to generate a particular number. L is the $\log_b N$ if $N = b^L$.

The **change-of-base rule** says that the log base b of a number N can be written in terms of another base, c, by dividing the log base c of N by the log base c of b. You change the base by dividing the log of the number by the log of the base.

CHECKPOINT

Rewrite in logarithmic form.

1. $2^7 = 128$ 2. $x^y = z$

Rewrite in exponential form.

3. $\log_3 81 = 4$ 4. $\log_r t = v$

5. Use your calculator and the change-of-base rule to find $\log_5 144$.

Properties of Logarithms

When you learned to work with exponents, there were three fundamental rules:

- To multiply powers of the same base, keep the base and add the exponents.

- To divide powers of the same base, keep the base and subtract the exponents.

- To raise a power to a power, keep the base and multiply the exponents.

Because logarithms are, in a sense, really exponents, it's not surprising that there are three corresponding fundamental rules for logs. In each, even though you won't see the base shown, you can assume the base of the logs is the same. The rules hold for any base, as long as it is the same throughout:

- To find the log of the product ab, add the $\log(a) + \log(b)$.

- To find the log of the quotient $\frac{a}{b}$, subtract $\log(a) - \log(b)$.

- To find the log of a power a^n, multiply $\log(a)$ times the exponent n.

Expanding an Expression

Let's look at a few examples of these rules in action. If you're asked to find $\log_{10}(3xy)$, you can apply the first rule and say $\log_{10}(3xy) = \log_{10}(3) + \log_{10}(x) + \log_{10}(y)$. In some situations, you might find it useful to think of $\log_5 250$ as $\log_5(125 \cdot 2)$, because that can become $\log_5(125 \cdot 2) = \log_5(125) + \log_5(2) = 3 + \log_5(2)$. The second rule tells you that $\log_3\left(\frac{x}{9}\right) = \log_3(x) - \log_3(9) = \log_3(x) - 2$, and the third rule is actually a variant of the first. If you are asked to find $\log_2(x^3)$, you could think of that as a product: $\log_2(x^3) = \log_2(x \cdot x \cdot x)$. That becomes $\log_2(x) + \log_2(x) + \log_2(x) = 3\log_2(x)$. The third rule is just a shortcut for this logic.

ALGEBRA TIP

If you see a logarithm written without a base shown, you can assume the base is 10.

The challenge is when two or more rules combine in the same problem. For instance, $\log\left(\frac{3x^2}{4y^3}\right)$ can first of all be seen as a quotient, so $\log\left(\frac{3x^2}{4y^3}\right)=\log\left(3x^2\right)-\log\left(4y^3\right)$. Each of those logs is the log of a product, so each can become the sum of logs.

$$\log\left(\tfrac{3x^2}{4y^3}\right)=\log\left(3x^2\right)-\log\left(4y^3\right)$$
$$=\log(3)+\log\left(x^2\right)-\left[\log(4)+\log\left(y^3\right)\right]$$

You can distribute the minus, and then deal with the powers.

$$\log\left(\tfrac{3x^2}{4y^3}\right)=\log\left(3x^2\right)-\log\left(4y^3\right)$$
$$=\log(3)+\log\left(x^2\right)-\left[\log(4)+\log\left(y^3\right)\right]$$
$$=\log(3)+\log\left(x^2\right)-\log(4)-\log\left(y^3\right)$$
$$=\log(3)+2\log(x)-\log(4)-3\log(y)$$

CHECKPOINT

Expand each logarithm to a sum, difference, and/or multiple of logs.

6. $\log_7(3y^5)$ 7. $\log_2(8n^4)$ 8. $\log_b\left(\frac{4x^2}{y}\right)$

9. $\log_3\left(x^2\sqrt{y}\right)$ 10. $\log\left(\frac{x^2-4}{x^2+3x-4}\right)$

Condensing an Expression

The rules work in both directions. You can use them to expand an expression, as shown previously, or to condense an expression involving several logs to a single logarithm. A sum of logs becomes the log of a product: $\log_b(x) + \log_b(y) = \log_b(xy)$. The difference of logs can be condensed to the log of a quotient, $\log_b(x)-\log_b(y)=\log_b\left(\frac{x}{y}\right)$, and a multiple, like $5\log_b(t)$, can be written as the log of a power, $\log_b(t^5)$. Using the rules, you can reduce an expression like $\log_b(8)+3\log_b(x)-2\log_b(y)-\tfrac{1}{2}\log_b(z)$ to a single log.

First, deal with exponents.

$$\log_b(8) + \log_b(x^3) - \log_b(y^2) - \log_b\left(z^{\frac{1}{2}}\right)$$

The sum of logs becomes the log of a product.

$$\log_b(8x^3) - \log_b(y^2) - \log_b\left(z^{\frac{1}{2}}\right)$$

Two terms are being subtracted. Let's factor out a minus.

$$\log_b(8x^3) - \left[\log_b(y^2) + \log_b\left(z^{\frac{1}{2}}\right)\right]$$

Now condense that sum, and as your last step, turn the difference of logs into the log of a quotient.

$$\log_b(8x^3) - \log_b\left(y^2 z^{\frac{1}{2}}\right) = \log_b\left(\frac{8x^3}{y^2 z^{\frac{1}{2}}}\right)$$

ALGEBRA TRAP

When condensing an expression involving logs, remember that your goal is a single log. The sum of two logs becomes the (single) log of a product, not the product of two logs. The difference of two logs becomes the (single) log of a quotient, not the quotient of logs.

CHECKPOINT

Condense each expression to a single log.

11. $3\log x + \frac{1}{2}\log y$

12. $8\log_3 t - 5\log_3 u$

13. $\frac{1}{2}\log(x+4) + 3\log(y+2)$

14. $\ln(x-5) + \ln(x+5) - 2\ln(x-1)$

15. $4\log_2 a + \frac{1}{2}\log_2 b - 3\log_2 c$

Solving Exponential Equations

An exponential function is one that contains a number, called the *base,* raised to a power that involves a variable. There may be a multiplier of the power, and there may be a constant term as well, but the function takes its name from that base and exponent structure. $f(x) = 2^x$ is a simple example of an exponential function, and a more complex example might look like $g(x) = -3\left(\frac{1}{2}\right)^{2x+5} + 4$.

Whenever you work with a function, you need to be able to solve equations involving that function, whether to find x-intercepts for the graph or to answer questions about the value of x that produces a certain function value. There are three common methods of solving equations involving exponentials, which I'll go over now.

Powers of the Same Base

If both sides of the equation can be written as powers of the same base, the *exponential equation* can translate to a simpler equation about the exponents themselves. So if $2^x = 2^3$, then $x = 3$. While most equations of this type are not quite so obvious, a little work will get you to a similar situation.

For example, if $-5 \cdot 3^{2x-5} = -\frac{5}{27}$, first divide by -5. You want to focus on the power; the multiplier is just a distraction: $-5 \cdot 3^{2x-5} = -\frac{5}{27}$ becomes $3^{2x-5} = \frac{1}{27}$.

 **DEFINITION**

An **exponential equation** is an equation containing one or more terms with a variable in the exponent position.

The left side is a power of 3, so you want to try to write the right side as a power of 3. You know that 27 is 3^3, so $\frac{1}{27} = \frac{1}{3^3} = 3^{-3}$. The equation becomes $3^{2x-5} = 3^{-3}$, and you can shift from that equation to the simpler $2x - 5 = -3$. Adding 5 tells you $2x = 2$, so $x = 1$.

Log of Both Sides

If it's not possible to write both sides as powers of the same base, you can turn to logarithms to help you solve exponential equations. If your equation is two equal powers, even if the bases are different, you can take the log of both sides using any convenient base. "Any convenient base" generally means one of the two bases that correspond to keys on your calculator, but you can use another base if you wish. Your calculator can produce common (base 10) logarithms or natural (base e) logs, but the change-of-base rule will allow you to find logs with any base you choose.

If $2^x = 3^{x-5}$, for example, you can't equate the exponents, but you can take the log of both sides.

$$\log(2^x) = \log(3^{x-5})$$

The log of a power can be rewritten; in this case, you can do that on both sides.

$$\log\left(2^x\right) = \log\left(3^{x-5}\right)$$
$$x\log\left(2\right) = \left(x-5\right)\log\left(3\right)$$

It may be tempting to reach for your calculator to find log(2) and log(3), but those will be decimals you'll need to round, and the calculation with them will be messy. Let's wait. Just remember that log(2) and log(3) are numbers, constants, and let's work with them as they are for now. Distribute log(3).

$$x\log(2) = x\log(3) - 5\log(3)$$

ALGEBRA TIP

The results of a calculation involving logs will be the same, whatever base you choose, as long you use the same base throughout the calculation. The choice of base may make your work simpler. $\ln(e^{5x}) = 5x$ is simpler to deal with than $\log(e^{5x}) = 5x\log(e)$.

Move all the terms that contain x to one side, and all terms without an x to the other. Factor out the x, and divide to isolate x.

$$x\log\left(2\right) - x\log\left(3\right) = -5\log\left(3\right)$$
$$x\left[\log\left(2\right) - \log\left(3\right)\right] = -5\log\left(3\right)$$
$$x = \frac{-5\log(3)}{\log(2)-\log(3)}$$

Now you can grab your calculator. The value of x is approximately 13.55.

You may need to do some algebra before you have two equal powers and can take the log of both sides. Simplify as much as you can, but if you cannot simplify to two equal powers or to a power equal to a constant, you may still be able to apply the log rules carefully to the equation.

You may find that you can rewrite a multiplier as a power of one of the bases already in your equation, and then multiply by adding exponents or divide by subtracting exponents, as seen in the following.

$$5 \cdot 7^{2x-1} = 81 \cdot 3^{x+4}$$

$$5 \cdot 7^{2x-1} = 3^4 \cdot 3^{x+4}$$

$$5 \cdot 7^{2x-1} = 3^{x+8}$$

Unfortunately, you can't eliminate the 5 that way, but don't give up yet. Take the log of both sides, and apply the rules for logs.

$$\log\left(5 \cdot 7^{2x-1}\right) = \log\left(3^{x+8}\right)$$

$$\log(5) + \log\left(7^{2x-1}\right) = \log\left(3^{x+8}\right)$$

$$\log(5) + (2x-1)\log(7) = (x+8)\log(3)$$

Distribute, and move all terms involving x to one side.

$$\log(5) + (2x-1)\log(7) = (x+8)\log(3)$$

$$\log(5) + 2x\log(7) - \log(7) = x\log(3) + 8\log(3)$$

$$2x\log(7) - x\log(3) = 8\log(3) + \log(7) - \log(5)$$

 **CALCULATOR CORNER**

Be sure to enclose denominators and complicated numerators in parentheses, and to close any parentheses your calculator automatically opened. Order of operations errors may occur if parentheses don't match properly.

Factor out the x and divide. The calculation won't be pretty, but it will get you a solution. Enclose both the numerator and the denominator in parentheses.

$$x\left[2\log(7) - \log(3)\right] = 8\log(3) + \log(7) - \log(5)$$

$$x = \frac{8\log(3) + \log(7) - \log(5)}{2\log(7) - \log(3)}$$

$$x \approx 3.267$$

CHECKPOINT

Solve by writing both sides as powers of the same base.

16. $2^{1-2x} = 4^{x+3}$

17. $81^{5x-7} = 27^{7x-11}$

Solve by taking the log of both sides.

18. $6^x = 5^{2x-1}$

19. $2(5^{3x}) = 41$

20. $1{,}500e^{0.06t} = 1{,}850$

Quadratic Form

There is a third type of exponential equation, which you may sometimes encounter, that is actually a *quadratic form*. For instance, the equation $5^{2x} - 7 \cdot 5^x + 12 = 0$ can be seen as a quadratic form if you write it as $(5^x)^2 - 7 \cdot (5^x) + 12 = 0$. You may even want to make a substitution, letting $a = 5^x$, so you can see the equation as $a^2 - 7a + 12 = 0$. Solve for a first.

$$a^2 - 7a + 12 = 0$$
$$(a-3)(a-4) = 0$$
$$a - 3 = 0 \qquad a - 4 = 0$$
$$a = 3 \qquad\qquad a = 4$$

 DEFINITION

A **quadratic form** is an equation that has the structure of a quadratic equation, even though it involves powers or expressions that do not fit the definition of a quadratic equation. The form can be made clearer by making a substitution of a single variable for a larger expression.

Next, step back and remember that $a = 5^x$, and solve the two little exponential equations, if possible.

$$a = 3 \qquad\qquad\qquad a = 4$$
$$5^x = 3 \qquad\qquad\qquad 5^x = 4$$
$$\log(5^x) = \log(3) \qquad \log(5^x) = \log(4)$$
$$x\log(5) = \log(3) \qquad x\log(5) = \log(4)$$
$$x = \tfrac{\log(3)}{\log(5)} \approx 0.683 \qquad x = \tfrac{\log(4)}{\log(5)} \approx 0.861$$

CHECKPOINT

Each of these equations is quadratic in form. Solve each equation for x.

21. $3^{2x} - 5(3^x) + 6 = 0$

22. $5^{4x} - 9(5^{2x}) + 20 = 0$

23. $\left(\frac{1}{2}\right)^{2x} - \left(\frac{1}{2}\right)^x - 30 = 0$

24. $e^{2x} - 4(e^x) + 4 = 0$

25. $9^x - 12(3^x) + 27 = 0$

Solving Log Equations

Equations involving logarithms may seem complex on first viewing, but if you focus on simplifying each side of the equation as completely as possible before beginning to solve, you'll find that they will reduce to one of two situations. As you simplify, remember that you want as few logs as possible, so apply rules for logs with the goal of reducing each side to a single log.

Log Equals Constant

If you're fortunate, you'll find that you have a single log on one side and a constant on the other, as in $\log_3(4x + 7) = 5$. In that situation, rewrite the equation as an exponential statement. Remember, the exponent you would put on 3 to make $4x + 7$ is 5. Once you've rewritten, you should be able to solve smoothly.

$$\log_3\left(4x+7\right) = 5$$
$$3^5 = 4x+7$$
$$243 = 4x+7$$
$$236 = 4x$$
$$59 = x$$

You may find that it takes a little bit of cleaning up before you have a single log equal to a constant. For instance, let's look at the equation $4\log_2\left(\frac{3x-1}{3}\right) = 20 - 4\log_2\left(3\right)$. Often there are many different ways you could rearrange an equation to try to simplify things, so it's good to keep a goal in mind. You want to see if you can get this to one log, so let's get the logs on the same side.

$$4\log_2\left(\tfrac{3x-1}{3}\right) = 20 - 4\log_2\left(3\right)$$
$$4\log_2\left(\tfrac{3x-1}{3}\right) + 4\log_2\left(3\right) = 20$$

Don't jump at moving those 4's up to the exponent position. You could, but notice that you can divide through the equation by 4.

$$4\log_2\left(\tfrac{3x-1}{3}\right) + 4\log_2\left(3\right) = 20$$
$$\log_2\left(\tfrac{3x-1}{3}\right) + \log_2\left(3\right) = 5$$

Don't grab your calculator to get a decimal approximation for the log base 2 of 3. Instead, combine the logs.

$$\log_2\left(\tfrac{3x-1}{3}\right) + \log_2\left(3\right) = 5$$
$$\log_2\left(\tfrac{3x-1}{3}\cdot 3\right) = 5$$
$$\log_2\left(3x-1\right) = 5$$

One log equals a constant, so rewrite as an exponential and solve.

$$\log_2\left(3x-1\right) = 5$$
$$2^5 = 3x-1$$
$$32 = 3x-1$$
$$33 = 3x$$
$$11 = x$$

Log Equals Log

If you weren't fortunate enough to have only one log, your simplifying should still have left you with a log on each side—two equal logs. To take a simple example first, if you know $\log_2 x = \log_2 5$, it isn't hard to see that $x = 5$. Technically, the method of solving $\log_2 x = \log_2 5$ is *exponentiation*. The theory says that if $\log_2 x = \log_2 5$, then $2^{\log_2 x} = 2^{\log_2 5}$. Now that may look worse than $\log_2 x = \log_2 5$, but if $\log_2 x$ is the exponent you put on 2 to make x and you put it on 2, you get x. If $\log_2 5$ is the exponent you put on 2 to get 5, then $2^{\log_2 5}$ is 5. So really, $2^{\log_2 x} = 2^{\log_2 5}$ just says $x = 5$. As a result, most people just say that if you have two equal logs, you "drop the log." For instance, if $\log_7(2x + 9) = \log_7(5x - 7)$, then $2x + 9 = 5x - 7$.

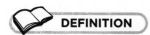 **DEFINITION**

> **Exponentiation** is using each side of an equation as an exponent on the same base, creating a new, equivalent equation of equal powers.

To solve the equation $\log_2(x^2 + 4x) = 3 + \log_2(x)$, you'll need to subtract $\log_2 x$ from both sides, and condense the difference of logs to the log of a quotient. You then convert to exponential form.

$$\log_2\left(x^2 + 4x\right) = 3 + \log_2\left(x\right)$$
$$\log_2\left(x^2 + 4x\right) - \log_2\left(x\right) = 3$$
$$\log_2\left(\tfrac{x^2+4x}{x}\right) = \log_2\left(x+4\right) = 3$$
$$2^3 = x + 4$$
$$8 - 4 = 4 = x$$

To solve $2\log(x + 3) = \log(x + 15)$, move the 2 up to its exponent position, and then drop the logs. Solve the resulting quadratic equation, and be sure to check your solutions. Remember, you cannot take the log of 0 or the log of a negative number, so extraneous solutions are not uncommon. Check your solution(s) in the original equation to be certain they don't leave you with the log of 0 or the log of a negative number.

$$2\log\left(x + 3\right) = \log\left(x + 15\right)$$
$$\log\left(x + 3\right)^2 = \log\left(x + 15\right)$$
$$\left(x + 3\right)^2 = x + 15$$
$$x^2 + 6x + 9 = x + 15$$
$$x^2 + 5x - 6 = 0$$
$$\left(x + 6\right)\left(x - 1\right) = 0$$
$$x + 6 = 0 \qquad x - 1 = 0$$
$$x = -6 \qquad \boxed{x = 1}$$
$$\textit{reject}$$

The first solution of $x = -6$ must be rejected, because if you substitute -6 for x, you have $2\log(-3) = \log(9)$ and you cannot find the log of -3. The other solution is acceptable, so $x = 1$.

CHECKPOINT

Solve each logarithmic equation.

26. $\log_5(2x - 5) = \log_5(x + 7)$

27. $\ln(x^2 - 8x + 5) = \ln(4 - 6x)$

28. $\log_2(6x + 5) = 5$

29. $\log_4(x) + \log_4(x - 3) = 1$

30. $\log(2x - 3) - \log(x + 1) = \log(x - 1)$

How to Graph Exponential and Logarithmic Functions

The graph of an exponential function has a distinctive shape: nearly flat on one side, and quite steep on the other. You can see in the following parent graph that as you move through the negative numbers, approaching $x = 0$, the function is increasing, but very slowly. Once you cross the y-axis, the rate of growth sharply increases.

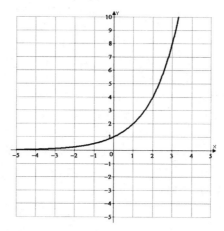

The domain of the exponential function is all real numbers, but the range is values of $y > 0$. The graphs of other exponential functions can be sketched using transformations, and those transformations may change the range, but not the domain.

To graph an exponential function, begin by noting the base of the power. The key points will vary depending on the base. The parent function, regardless of base, has a y-intercept of $(0, 1)$, but the other key points will be $(1, \text{base})$ and $(-1, \frac{1}{\text{base}})$. For the preceding parent graph, $y = 2^x$, the key points would be $(0, 1)$, $(1, 2)$, and $(-1, \frac{1}{2})$.

Translations of the parent graph show up in the equation as additions and subtractions. For example, $y = 2^{x-3} + 1$ is the parent graph moved 3 units right and 1 unit up, as you can see in the following.

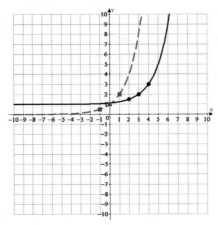

Because of the shape of the exponential graph and the fact that different bases are possible, stretches can be difficult to recognize from just looking at the graph, but as always, they show up in the equation as multipliers. For instance, $y = 5 \cdot 2^x$ has a vertical stretch by a factor of 5. Reflecting the exponential parent graph across the y-axis produces a graph that decreases rather than increases.

ALGEBRA TRAP

Remember your order of operations. Exponents have a higher priority than multiplication, so if you have a multiplier—like the 5 in $y = 5 \cdot 2^x$—in front of an exponential expression, remember that you cannot combine the multiplier with the base of the power. So $5 \cdot 2^x \neq 10^x$.

This is because the change to the equation that creates the reflection—multiplying x by –1— changes the expression from one with a positive exponent to one with a negative exponent. When $y = 2^x$ becomes $y = 2^{-x}$, you're actually looking at $y = \left(2^{-1}\right)^x = \left(\frac{1}{2}\right)^x$, and a base less than 1 produces smaller y-values as x increases, shown as follows.

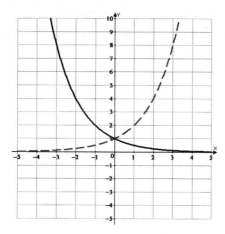

A logarithmic function is the inverse of an exponential function. Its domain is all real numbers greater than 0, and its range is all reals. The parent graph for the exponential function, shown as follows, is the reflection of the exponential parent, over the line $y = x$. The key points of the exponential parent graph $y = 2^x$ are $(0, 1)$, $(1, 2)$, and $(-1, \frac{1}{2})$. The key points of its inverse, $y = \log_2(x)$, are $(1, 0)$, $(2, 1)$, and $(\frac{1}{2}, -1)$.

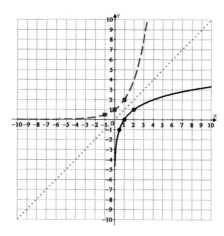

The parent graph can be transformed to produce the graphs of other logarithmic functions. For example, the following graph of $g(x) = \log_3(x + 2) - 4$ is a transformation of $y = \log_3(x)$, moved 2 units left and 4 down.

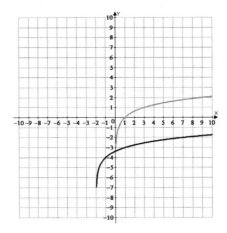

The following graph of $f(x) = 4 - 3\log_2(x + 5)$ shifts the parent graph 5 units left, stretches it by a factor of 3, reflects it over the x-axis, and moves it up 4 units.

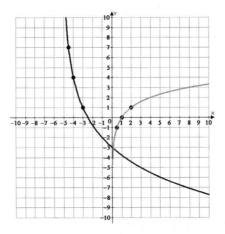

CHECKPOINT

Sketch a graph of each function. Draw as much of the graph as you can before using a graphing calculator or software.

31. $f(x) = 2^{x-3} + 5$

32. $g(x) = -\frac{1}{2}\left(3^{x+4}\right) - 2$

33. $f(x) = 3\log_2(x - 1)$

34. $g(x) = 5 - \log_3(x + 2)$

35. $f(x) = \ln(4 - x) + 3$

Growth and Decay

Exponential equations model many circumstances of growth or decay, increase or decrease. When interest—whether interest earned on an investment or interest paid on a loan— is compounded, the total amount in the account is given by the exponential equation $A = P\left(1 + \frac{r}{n}\right)^{nt}$, where P is the original amount, known as the principal; A is the total amount; r is the rate of interest; n is the number of times per year the interest is compounded; and t the number of years. So a sum of \$10,000 invested for 10 years at 6% per year, compounded quarterly, will produce $A = 10,000\left(1 + \frac{0.06}{4}\right)^{4 \cdot 10} = 10,000\left(1.015\right)^{40} \approx \$18,140.18$.

As compounding occurs more often, and n becomes large, this formula approaches what's called the continuous compounding model, $A = Pe^{rt}$, where P is the principal, A the total amount, r is the rate, and t the time in years. The constant e is approximately 2.71828.... This equation, often called the *Pert formula*, can be used in many situations of exponential growth or decay. If a certain bacterium doubles every 12 minutes, for example, you can determine its growth rate by substituting $P = 1$, $A = 2$, and $t = 12$, and then solving for r.

 **DEFINITION**

The **Pert formula** is a model for continuous growth or decay, based on the original amount, P; the rate of growth or decay, r; and the time, t. For growth, the rate is positive; for decay, it is negative.

$$A = Pe^{rt}$$
$$2 = 1 \cdot e^{12r}$$
$$\ln(2) = 12r \ln(e) = 12r$$
$$\frac{\ln(2)}{12} = r$$
$$r \approx 0.0578$$

Once you have the rate of growth per minute, you can use $A = Pe^{0.0578t}$ to find the number of bacteria after a certain time, or the length of time necessary to reach a certain population.

For instance, if a radioactive element has a half-life of 20 years, you can use a Pert model to find its rate of decay.

$$A = Pe^{rt}$$
$$\tfrac{1}{2} = 1 \cdot e^{r \cdot 20}$$
$$\ln\left(\tfrac{1}{2}\right) = 20r \ln(e) = 20r$$
$$r = \frac{\ln\left(\frac{1}{2}\right)}{20} \approx -0.0347$$

The rate is a negative number because it is a rate of decay, rather than rate of growth. If you want to know how long this element must decay before there is less than 1% left, use $A = Pe^{rt}$ with $A = 0.01$, $P = 1$, and $r = -0.0347$.

$$A = Pe^{rt}$$

$$0.01 = 1 \cdot e^{-0.0347t}$$

$$\ln(0.01) = -0.0347t \ln(e) = -0.0347t$$

$$t \approx \frac{\ln(0.01)}{-0.0347}$$

$$t \approx 132.713$$

The level of this element will drop below 1% after about 133 years.

CHECKPOINT

Use the given formula to solve each problem.

36. How much money must you deposit in an account that pays 4% interest, compounded quarterly, if you want to have $10,000 in the account 18 years later? (Use $A = P\left(1 + \frac{r}{n}\right)^{nt}$.)

37. How long will it take to double your money if it is invested at an annual interest rate of 3.2% compounded continuously? (Use $A = Pe^{rt}$.)

38. A colony of 100 bacteria is provided with necessary nourishment and allowed to reproduce. After 48 hours, the colony contains 500 bacteria. Find the growth rate and use it to predict when the population of the colony will reach 1,000,000. (Use $A = Pe^{rt}$.)

39. Pu-240, an isotope of Plutonium, has a half-life of 6,560 years. (A quantity of Pu-240 will slowly decay over 6,560 years to half of the original amount.) Find the rate of decay, and use it to estimate how much Pu-240 will remain in 2050 from 10 grams measured in 2015. (Use $A = Pe^{rt}$.)

40. The pH of a substance measures how acidic or basic the substance is. The pH of distilled water is 7. Acids have pH less than 7 and bases greater than 7. The formula pH = $-\log(H^+)$ defines the pH of a substance as the negative log (base 10) of the hydrogen ions in the solution. Find the hydrogen ions in a solution of lemon juice if its pH is approximately 2.1.

The Least You Need to Know

- A logarithm with a base of b of a number N is the exponent you would put on b to make N.

- To multiply powers of the same base, keep the base and add the exponents. To take the log of a product, add the logs of the factors. To divide powers, subtract the exponents. To take the log of a quotient, subtract log(numerator) − log(denominator). Log of a power equals exponent times log of base, because you raise a power to a power by multiplying exponents.

- Solve an exponential equation by writing both sides as powers of the same base and equating the exponents, or by taking the log of both sides.

- Solve a log equation by rewriting it as an exponential or by simplifying it as two equal logs and dropping the logs.

- Because exponential and logarithmic functions are inverses, their graphs are reflections over the line $y = x$.

- Use exponential and logarithmic functions to solve problems about growth and decay.

Probability

Did you ever buy a lottery ticket? Did you ever read the information about your chance of winning? Most people don't, and so have no idea how small their chance of winning the top prize actually is. Too many people, asked what their chances are, would answer "50–50," better than 50–50, or worse than 50–50. And many of those people don't really know what "50–50" means.

In this chapter, you get a better grasp of what probability really means and how it's calculated. Part of that calculation involves counting—sometimes up to very large numbers—so you learn some quick methods for getting those counts and using them in calculating probabilities. If you stay for all of that, the bonus at the end is a quicker method of raising a binomial to a power, thanks to some of the counting techniques you learned.

Basic Counting Principle

You learned to count as a child, and you might think that's all there is to it. Consider this situation, however. You're buying a brand-new car, straight from the factory, with exactly the options you want. You have to choose the make, the model, and the color (exterior and interior). Also, do you want two-door or four-door, leather or fabric? You can choose automatic or manual transmission and rear-wheel drive, front-wheel drive, or all-wheel drive. Do you want the sun roof? GPS? Bluetooth? How many cars do you have to choose from? Did you count?

A situation like that is a far cry from the counting you did as a child, and thank goodness, there's a way to organize your thinking about it. You can do a visual image of your decision making with a *tree diagram*. Here's a simplified example. You need to choose whether you want a sedan, wagon, or SUV, so the tree starts with three branches, as shown in the following figure.

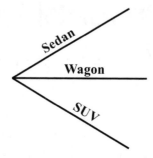

You then need to decide on color: silver, black, or red. You can have the sedan and the wagon in any of those three colors, but the SUV only comes in silver or black. Each branch of the tree grows branches.

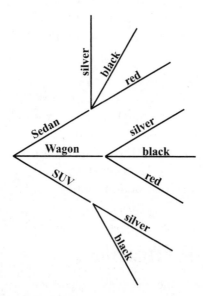

The process continues with new branches for each choice to be made, but as you can see, the drawing of the tree could quickly become unwieldy. You could run out of space to draw before you finished all your branches and might never be able to count how many different end results are available.

The *basic counting principle* gives you a more efficient way to get that count. Think about how many options you have for each choice. Automatic or manual transmission? Two options. Front-wheel, rear-wheel, or all-wheel drive? Three options. Sun roof—yes or no? Two options. The basic counting principle tells you that if you have two options for the first choice, three for the second, and two for the third, you have a total of $2 \cdot 3 \cdot 2 = 12$ vehicles to choose from.

DEFINITION

A **tree diagram** is a visual device to show the many different ways a set of choices can be made. Each branch, followed to its tip, represents one complete set of choices. The **basic counting principle** says the number of different ways a series of choices can be made is the product of the number of different options for each choice.

Considering another example, if you order your pizza with 2 toppings, and there are 12 choices for toppings and 3 different crust styles, the number of different pizzas you could order is $\underbrace{12}_{\text{topping 1}} \cdot \underbrace{12}_{\text{topping 2}} \cdot \underbrace{3}_{\text{crust}} = 432$. Of course, if you don't want double anchovies or double pepperoni, it will only be $\underbrace{12}_{\text{topping 1}} \cdot \underbrace{11}_{\text{topping 2}} \cdot \underbrace{3}_{\text{crust}} = 396$.

CHECKPOINT

Use the basic counting principle to calculate the number of different outcomes in each situation. Draw a tree diagram if you find it helpful.

1. A lottery requires you to choose three single digits, and you are permitted to repeat digits. How many different entries are possible?

2. If the lottery in the previous question bans repeated digits, how many different entries are possible?

3. In New Jersey, standard-issue license plates contain two groups of three characters each, a total of six characters. Each character can be one of the 26 letters of the alphabet or one of the 10 digits. How many standard license plates can be created?

4. Phone numbers in the United States have a three-digit area code, a three-digit exchange, and a four-digit number (example: 123-555-1234). There are, in reality, some restrictions on what numbers may be used, but if those restrictions are ignored, how many phone numbers are theoretically possible?

5. A large national lottery requires players to choose 5 numbers from a group of 59 without repeats, and a single number from a group of 35. (It is permissible for the single number to duplicate one of the set of five numbers.) How many different entries are possible?

Combinations and Permutations

If you play on a basketball team with seven players, and after practice you line up at the vending machine to get something cool and wet to replenish all the fluid you sweat out, how many different ways can your team line up? Basically, that's a counting principle question. There are seven people who could be first, six who could be second, five who could be third, and so on, down to one person left to be last. The product $7 \cdot 6 \cdot 5 \cdot 4 \cdot 3 \cdot 2 \cdot 1 = 5{,}040$ is the number of ways your team can line up. (So how is it that you're never first?) That product also has a name: 7 *factorial*, written 7! The product of the integers from a positive integer n, down to 1, is $n!$ That product is also the number of *permutations* of seven people taken seven at a time. A permutation is an order or arrangement, and you calculated there were 5,040 orders in which your team could line up. The permutations of n (7) things taken n (7) at a time is $n!$ (7!).

Now what if your team was lining up to shoot free throws, but three people had to retrieve balls, so only four could line up to practice? That's still a permutation—an arrangement—but it won't be a factorial, because there will only be four of the seven players in the line. There are seven players who could be first in line, six who could be second, five who could be third, and four who could be the fourth. That's $7 \cdot 6 \cdot 5 \cdot 4 = 840$ different orders. The permutations of n things taken r at a time can be found as you just did, by using the basic counting principle, or with the formula $_nP_r = \frac{n!}{(n-r)!}$. The symbol $_nP_r$ represents the permutations of n things taken r at a time. In the example of the free throws, you would write $_7P_4$, the number of arrangements of seven players lining up four at a time. The formula says $_7P_4 = \frac{7!}{(7-4)!} = \frac{7!}{3!} = \frac{7 \cdot 6 \cdot 5 \cdot 4 \cdot \cancel{3 \cdot 2 \cdot 1}}{\cancel{3 \cdot 2 \cdot 1}} = 7 \cdot 6 \cdot 5 \cdot 4 = 840$. It starts with 7!, but the extra factorial in the denominator cancels out the last three factors, leaving only the four positions to fill in the line.

What if you were choosing your starting five from the seven members of the team? Although players play different positions—guard, center, and so on—the order in which you're chosen to start doesn't really matter. A starting five of Smith, Jones, Rodriguez, Johnson, and Hong is the same as Johnson, Smith, Rodriguez, Hong, and Jones. The order isn't significant in this situation. When order doesn't matter, you're looking for the *combinations*, not the permutations, of seven things taken four at a time. A combination is only interested in the number of different groups, not the order or arrangement of the members of the group. The formula for the combinations of n things taken r at a time is $_nC_r = \frac{n!}{(n-r)!r!}$. It looks a lot like the permutations formula, but it has an extra factor in the denominator. Let's look at a small example to see why.

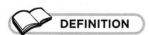 **DEFINITION**

> A **factorial** is the product of positive integers in sequence. If n is a positive integer, n factorial (or $n!$) is the product of positive whole numbers from 1 to n. A **permutation** is an order or arrangement. The digits 2, 5, and 8 can be arranged in six different orders—258, 285, 528, 582, 825, and 852—meaning there are six permutations of those three digits. A **combination** is a collection of objects for which order is not significant.

Take three letters: *A*, *B*, and *C*. The permutations of those three letters taken two at a time are a small-enough set that you can list them: *AB*, *AC*, *BA*, *BC*, *CA*, *CB*. There are six permutations of three things taken two at a time. If you're only interested in combinations, however, *AB* and *BA* are the same combination. *BC* and *CB* are the same, and *CA* and *AC* are duplicates; each combination has two permutations. There are six permutations but only three combinations.

The *r*! in the denominator of the combinations formula reduces the count by eliminating different arrangements of the same combination. In choosing a group of five, there are $_7P_5 = \frac{7!}{(7-5)!} = \frac{7!}{2!} = \frac{7 \cdot 6 \cdot 5 \cdot 4 \cdot 3 \cdot \cancel{2 \cdot 1}}{\cancel{2 \cdot 1}} = 2{,}520$ permutations but far fewer combinations. A group of five can be arranged in 5! = 120 ways. The 2,520 permutations are 21 groups of 120 arrangements. The combination formula breaks that down for you:

$$_7C_5 = \frac{7!}{(7-5)!5!} = \frac{7 \cdot 6 \cdot 5 \cdot 4 \cdot 3 \cdot \cancel{2 \cdot 1}}{\cancel{(2 \cdot 1)} \cdot (5 \cdot 4 \cdot 3 \cdot 2 \cdot 1)} = \frac{7 \cdot 6 \cdot \cancel{5 \cdot 4 \cdot 3}}{\cancel{5 \cdot 4 \cdot 3} \cdot 2 \cdot 1} = \frac{7 \cdot 6}{2 \cdot 1} = 21.$$

 CALCULATOR CORNER

Your graphing calculator probably has functions for factorial, permutation, and combination. These may be hidden away in a menu rather than assigned to individual keys. Depending on the calculator, you may enter the permutations of five things taken two at a time as 5, the permutations function, and then 2, or as the permutation function followed by (5, 2).

 CHECKPOINT

Find the number of permutations or combinations as indicated.

6. $_4P_3$

7. $_6P_2$

8. $_{10}C_4$

9. $_7C_3$

10. A department of 14 professors must select a committee of 5 of their members to represent them in salary negotiations. If all the members of the committee have equal standing, how many different committees are possible?

Pascal's Triangle

There is a device that helps to find the number of combinations quickly. *Pascal's triangle* is an arrangement of numbers that can be generated by a simple addition pattern. Start with a 1 at the top. Underneath, add a new row with a 1 to start and a 1 to end. The next row starts with a 1 and will end with a 1, but in between, will have the sum of the two 1s above. The following figure shows you how this looks.

DEFINITION

Pascal's triangle is a triangle arrangement of numbers that can be generated by starting and ending each row with a 1 and filling the row with sums of pairs of elements in the rows above. The triangle contains the number of combinations of n things taken 0, 1, ..., n at a time, which are the coefficients necessary to form powers of binomials.

Each succeeding row is formed by starting with a 1, generating new elements by adding the two immediately above as you work across, and ending with a 1.

Pascal's triangle is more than a novelty, however. It contains a chart of the combinations of n things taken r at a time, for any values of n and r you choose. For example, if you are looking for the number of combinations of four things taken two at a time, find the row that begins with 1 and 4. That row has the combinations of 4 things taken 0 at a time, 1 at a time, 2 at a time, 3 at a time, and 4 at a time. The first 1 is the combinations of 4 things taken 1 at a time. The first 4 is the combinations of 4 things taken 1 at a time. The combinations of 4 things, taken 2 at a time, are represented by the 6. The elements in the row are 1, 4, 6, 4, 1: $_4C_0 = 1$, $_4C_1 = 4$, $_4C_2 = 6$, $_4C_3 = 4$, and $_4C_4 = 1$.

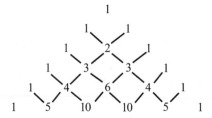

The combinations of 5 things taken 3 at a time can be found by looking at the row beginning with 1 and 5, and counting out that $_5C_0 = 1$, $_5C_1 = 5$, $_5C_2 = 10$, and $_5C_3 = 10$. There are 10 combinations of 5 things taken 3 at a time.

CHECKPOINT

Use Pascal's triangle to find the number of combinations.

11. $_5C_4$ 12. $_3C_2$ 13. $_6C_3$ 14. $_4C_4$ 15. $_8C_1$

Calculating Probability

The reason you want a quick way to count is that the definition of probability is based on counting things. When you conduct an experiment or make observations, there are usually many possible things that can happen—many outcomes. If a particular outcome you are interested in occurs, you consider that a success. The probability of an event E is equal to the number of ways the event can happen divided by the number of things that can happen.

$$P(E) = \frac{\text{\# successes}}{\text{\# outcomes}}$$

To make any sense of that definition, you'll need to be able to count the number of possible outcomes and the number of successes. For example, the probability of drawing a red seven from a standard deck of playing cards is equal to the number of red sevens, which is 2, over the number of different cards you could draw, which is 52.

$$P(\text{red } 7) = \tfrac{2}{52} = \tfrac{1}{26}$$

The probability of an event is always a number between 0 and 1. If the probability of an event is 0, there are no successes. The event did not happen and cannot happen; it's impossible. If the probability is 1, the number of successes is equal to the number of outcomes. All the outcomes are successes. That event always happens; it's certain.

ALGEBRA TRAP

Probabilities of events can only be combined if they come from the same universe. For example, it would make no sense to add your chance of choosing to wear your brown shoes to your chance of winning the lottery. While it's possible to talk about probabilities of one type of event happening after another type, you can't add the probabilities from different experiments.

Say you have 8 take-out menus in your desk at work. Two of the menus are for burger joints; three are Chinese restaurants; and the other three are pizza, sushi, and Indian. Each day you choose one menu at random and order from that shop. The probabilities are listed in the following table.

Type	Burgers	Chinese	Pizza	Sushi	Indian
Probability	$\frac{2}{8} = \frac{1}{4}$	$\frac{3}{8}$	$\frac{1}{8}$	$\frac{1}{8}$	$\frac{1}{8}$

Notice that the total of all the probabilities is 1. If you add $\frac{2}{8} + \frac{3}{8} + \frac{1}{8} + \frac{1}{8} + \frac{1}{8}$, the total is 1. That's true for any listing of all the possible outcomes, because every possibility is listed, so one of them is certain to happen. A *probability distribution* is a listing of all possible outcomes and the probability of each. The sum of all the probabilities in a distribution is 1.

"What's the probability I'll have pizza today?" is not a question that will hold your attention very long, however. The more interesting questions are combinations of the simple ones, and the most common ones are questions of the form "What is the probability of A or B?" and "What is the probability of A and B?" The question that can be challenging but necessary is the question about *conditional probability*: "What is the probability that B will happen, if I know that A has happened?"

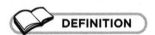

DEFINITION

A **probability distribution** is a list of all the possible outcomes and the probability of each. The **conditional probability** of an event is the probability the event occurs, given that another event has already occurred. The probability of B, given A, is $P(B|A) = \frac{P(A \text{ and } B)}{P(A)}$. If the events are independent, this will simply be the probability of the first event times the probability of the second.

Dependent vs. Independent Events

Two events, A and B, are dependent if A happening changes the probability of B. If you pull one card from a standard deck of playing cards, the probability of drawing an ace is $\frac{4}{52} = \frac{1}{13}$. If you draw one card, look at it, replace the card, and then draw again, the probability of drawing an ace on the second draw is still $\frac{4}{52} = \frac{1}{13}$. Replacing the card assured that the probabilities will be the same on the second draw. If you draw a card and don't replace the first card before drawing again, however, the probability of an ace on the second draw changes depending on what the first card was. If you draw an ace on the first pick and don't put it back, there are only three aces left in the deck, and only 51 cards in total to draw from: P (ace on the second draw, given an ace on the first) $= \frac{3}{51} = \frac{1}{17}$. If the first card drawn was not an ace, there are four aces left out of the 51: P (ace on the second draw, given no ace on the first) $= \frac{4}{51}$.

For independent events, $P(A \text{ and } B) = P(A) \cdot P(B)$. For instance, the probability you would randomly select pizza two days in a row is $P(\text{pizza and pizza}) = \frac{1}{8} \cdot \frac{1}{8} = \frac{1}{64}$. There are 8 choices the first day and 8 choices the second day, so you have a total of 64 ways you can randomly choose lunch on two consecutive days. Only one of them includes having pizza both days.

ALGEBRA TIP

If two events are independent, the probability $P(A \text{ and } B) = P(\text{A}) \cdot P(B)$. If the two events are dependent, the probability of both happening will still be a product, but the probabilities being multiplied will change: $P(A \text{ and } B) = P(A) \cdot P(B|A)$.

Disjoint vs. Connected Events

Two events are disjoint, or mutually exclusive, if they cannot happen at the same time. For example, an animal selected at random from a zoo cannot be both a lion and an elephant. If A and B are disjoint, $P(A \text{ or } B) = P(A) + P(B)$. In the case of the animal, $P(\text{lion or elephant}) = P(\text{lion}) + P(\text{elephant})$. As another example, assuming you only eat one lunch in a day, the probability you randomly select Chinese food or sushi is $P(\text{Chinese}) + P(\text{sushi}) = \frac{3}{8} + \frac{1}{8} = \frac{4}{8} = \frac{1}{2}$.

Back at the zoo, if you ask for the probability of randomly choosing a white animal or a polar bear, those are not disjoint. When you count white animals, you are also counting polar bears along with other white animals; they're groups with elements in common. If you calculate $P(\text{white animal})$ and $P(\text{polar bear})$, the polar bears will be counted in both groups, so your count of successes will be too high. When A and B are overlapping like this, $P(A \text{ or } B) = P(A) + P(B) - P(A \text{ and } B)$. So for the probability involving a white animal and a polar bear, $P(\text{white animal or polar bear}) = P(\text{white animal}) + P(\text{polar bear}) - P(\text{white animal and polar bear})$. That subtraction adjusts for the double counting. Let's use this formula with another example. If you draw a card at random from a standard deck of playing cards, $P(\text{red card}) = \frac{1}{2}$ and $P(\text{ace}) = \frac{1}{13}$, but the red card and the ace are not disjoint. The deck contains two cards that are red aces. They are counted in the 26 of 52 red cards and counted again in the 4 of 52 aces. So the probability you randomly select a red card or an ace is $P(\text{red or ace}) = P(\text{red}) + P(\text{ace}) - P(\text{red and ace}) = \frac{1}{2} + \frac{1}{13} - \frac{2}{52} = \frac{26}{52} + \frac{4}{52} - \frac{2}{52} = \frac{28}{52} = \frac{7}{13}$.

CHECKPOINT

Find the indicated probability.

16. If a penny and a dime are both fair coins and both are tossed, what is the probability they will land showing a head on the penny and a tail on the dime?

17. A bag contains seven blue candies, four red candies, three orange candies, and one yellow candy. If you draw a candy at random, what is the probability it will be red or orange?

18. In the situation described in the previous question, if you draw a candy, do not replace it, and then draw a second, what is the probability your second candy is blue, given that the first is orange?

19. If you draw two candies from the bag described previously without replacement, what is the probability you draw a blue candy on the first pick and a yellow on the second?

20. Researchers have found that, if you are exposed to a particular virus, there is a 45% chance you will contract the disease. If you contract the disease, there is a 10% chance your symptoms will be severe enough to require hospitalization. What is the probability you will contract the disease and require hospitalization?

Binomial Theorem

All this talk about counting does come around to help in algebra. You probably know, by memorization or by experience, that the square of a binomial like $(a + b)$ is $(a + b)^2 = a^2 + 2ab + b^2$. Squaring $3x - 5$ produces $(3x - 5)^2 = (3x)^2 + 2(3x)(-5) + (-5)^2$, which simplifies to $9x^2 - 30x + 25$. Of course, you could just FOIL, but once you've done it a number of times, you can probably anticipate the pattern.

To cube a binomial, you might use the pattern of the perfect square trinomial to get part of the work done, and then multiply by the third trinomial.

$$(x-2)^3 = (x-2)(x^2 - 4x + 4)$$
$$= x^3 - 4x^2 + 4x - 2x^2 + 8x - 8$$
$$= x^3 - 6x^2 + 12x - 8$$

You may also remember the pattern of a cube of a binomial: $(a + b)^3 = a^3 + 3a^2b + 3ab^2 + b^3$. If you raise $x + 4$ to the third power, you get the following.

$$(x+4)^3 = x^3 + 3x^2(4) + 3x(4)^2 + 4^3$$
$$= x^3 + 12x^2 + 48x + 64$$

If you were asked to raise a binomial to the fourth or fifth power, however, you might be reluctant, because you don't have a memorized form for powers greater than 3. You'll face a lot of multiplying and collecting like terms. Now that you're acquainted with combinations and Pascal's triangle, however, there is a pattern you can use to raise a binomial to any power quickly.

The *binomial theorem* tells you that, for any binomial, the nth power of the binomial is equal to the following.

$$(a+b)^n = \binom{n}{0}a^n b^0 + \binom{n}{1}a^{n-1}b^1 + \binom{n}{2}a^{n-2}b^2 + \ldots + \binom{n}{n-2}a^2 b^{n-2} + \binom{n}{n-1}a^1 b^{n-1} + \binom{n}{n}a^0 b^n$$

 **DEFINITION**

The **binomial theorem** shows the pattern of terms of the polynomial produced when a binomial is raised to a power:
$$(a+b)^n = a^n + \binom{n}{1}a^{n-1}b + \binom{n}{2}a^{n-2}b^2 + \ldots + \binom{n}{n-2}a^2 b^{n-2} + \binom{n}{n-1}ab^{n-1} + \binom{n}{n}b^n.$$

Okay, admittedly that doesn't look like a better method at first glance, but let's have a closer look. Each term begins with a number of combinations of n things. Remember, n is the power to which you're raising the binomial. The first term begins with the combinations of n things taken 0 at a time. That's always 1. The next term has n things taken 1 at a time, the third term n things taken 2 at a time, and so on. You can pull those numbers from Pascal's triangle.

Each term has a power of a times a power of b. The first term has a to the nth power and b to the 0 power. The powers of a go down and the powers of b go up, and the two exponents always add to n. Let's look at some examples.

$$(x+3)^4 = \binom{4}{0}x^4(3)^0 + \binom{4}{1}x^3(3)^1 + \binom{4}{2}x^2(3)^2 + \binom{4}{3}x^1(3)^3 + \binom{4}{4}x^0(3)^4$$

The preceding shows that the coefficients are the combinations of 4 things, taken 0 at a time through 4 at a time. Read them right across the appropriate row of Pascal's triangle.

$$(x + 3)^4 = 1x^4(3)^0 + 4x^3(3)^1 + 6x^2(3)^2 + 4x^1(3)^3 + 1x^0(3)^4$$

Next, you can evaluate the powers of 3.

$$(x + 3)^4 = 1x^4(1) + 4x^3(3) + 6x^2(9) + 4x^1(27) + 1x^0(81)$$

Simplify each term.

$$(x + 3)^4 = x^4 + 12x^3 + 54x^2 + 108x + 81$$

The pattern may look intimidating at first, but this is faster and easier than multiplying the binomials.

Here's another example. To raise $2x - 1$ to the sixth power, begin by setting up the pattern.

$$(2x-1)^6 = \binom{6}{0}(2x)^6(-1)^0 + \binom{6}{1}(2x)^5(-1)^1 + \binom{6}{2}(2x)^4(-1)^2 + \binom{6}{3}(2x)^3(-1)^3 + \binom{6}{4}(2x)^2(-1)^4$$
$$+ \binom{6}{5}(2x)^1(-1)^5 + \binom{6}{6}(2x)^0(-1)^6$$

Use Pascal's triangle for the coefficients.

$$
\begin{array}{ccccccccccccc}
&&&&&& 1 &&&&&& \\
&&&&& 1 && 1 &&&&& \\
&&&& 1 && 2 && 1 &&&& \\
&&& 1 && 3 && 3 && 1 &&& \\
&& 1 && 4 && 6 && 4 && 1 && \\
& 1 && 5 && 10 && 10 && 5 && 1 & \\
\boxed{1} && \boxed{6} && \boxed{15} && \boxed{20} && \boxed{15} && \boxed{6} && \boxed{1} \\
\end{array}
$$

$$(2x-1)^6 = 1(2x)^6(-1)^0 + 6(2x)^5(-1)^1 + 15(2x)^4(-1)^2 + 20(2x)^3(-1)^3 + 15(2x)^2(-1)^4$$
$$+ 6(2x)^1(-1)^5 + 1(2x)^0(-1)^6$$

Evaluate the powers of -1.

$$(2x-1)^6 = 1(2x)^6(1) + 6(2x)^5(-1) + 15(2x)^4(1) + 20(2x)^3(-1) + 15(2x)^2(1)$$
$$+ 6(2x)^1(-1) + 1(2x)^0(1)$$

Evaluate the powers of $2x$.

$$(2x-1)^6 = 1(64x^6)(1) + 6(32x^5)(-1) + 15(16x^4)(1) + 20(8x^3)(-1) + 15(4x^2)(1)$$
$$+ 6(2x)(-1) + 1(1)(1)$$

Simplify each term.

$$(2x - 1)^6 = 64x^6 - 192x^5 + 240x^4 - 160x^3 + 60x^2 - 12x + 1$$

CHECKPOINT

Use the binomial theorem and Pascal's triangle to expand each power of a binomial.

21. $(x - 2)^3$ 22. $(x + 3)^4$ 23. $(2x + 1)^5$

24. $(3x - 2)^4$ 25. $(x + 1)^7$

The Least You Need to Know

- The basic counting principle says the number of ways in which a series of choices can be made is the product of the number of options for each choice.

- The $n!$ symbol denotes the factorial or product of the whole numbers from n down to 1.

- A permutation is an arrangement of a group of objects for which order matters. The number of permutations of n things taken r at a time is $_nP_r = \frac{n!}{(n-r)!}$.

- A combination is a group in which order does not matter. The number of combinations of n things take r at a time is $_nC_r = \frac{n!}{r!(n-r)!}$. Pascal's triangle shows the number of combinations of n things taken 0 through n at a time across its rows.

- The probability of an event is a number between 0 and 1 that is equal to the number of ways the event can occur divided by the total number of possible outcomes. The probability of two events both occurring is $P(A \text{ and } B) = P(A) \cdot P(B)$. The probability either of two events will occur is $P(A \text{ or } B) = P(A) + P(B) - P(A \text{ and } B)$. The conditional probability of B given A is $P(B|A) = \frac{P(A \text{ and } B)}{P(A)}$.

- The binomial theorem provides a quick method of writing the expanded form of a power of a binomial.

Extra Practice

Is there any job in the world that doesn't require paperwork? It seems there are always reports to file and forms to fill out, even if the main part of your job is service or crafting or art. Algebra comes with its own paperwork, and that's what the final part of this book provides.

Like all branches of mathematics, algebra includes both concepts and skills—and skills, in particular, need to be reinforced by practice. This part is only one chapter, but it's a chapter that provides extra practice on the skills set out in every chapter that precedes it. Use this part to clarify skills as you learn them or to reinforce or review them later on. Answers for these problems can be found in Appendix D, so you can check your work.

Extra Practice Questions

Chapter 1

Place each number in all the sets to which it belongs: natural, integer, rational, irrational, real.

1. 6.725

2. −92

3. $\sqrt{182}$

4. −5

5. $\frac{42}{5}$

Simplify each expression.

6. $9(2 − 5) \div 3 + 7 − 12$

7. $5a − 12b \div 4 − 8b + 12a$

8. $15(2x + 4) − 25x + 3(1 − 10y)$

9. $\frac{18x}{-2x} - \frac{24x+81}{3}$

10. $\frac{6\left(12t^2 - 15t\right)}{9t}$

Solve each equation or inequality, simplifying before solving.

11. $5x + 3(4x + 9) = 13 − 8(4 − 5x)$

12. $12 − (17x + 15) = 27 − 13x$

13. $5(6x + 3) = 8 − (4x + 15) + 2x$

14. $1 − 4(x − 2) < 2 + 2x − 9$

15. $3x − (12 − 5x) \geq 9 + 6x − 21$

Graph each line.

16. $y = -\frac{2}{3}x + 5$

17. $y = 4x − 7$

18. $3x − 2y = 12$

19. $7x + 3y = −21$

20. $4y − 22 = 11x$

Find the equation of the line described.

21. slope = $\frac{2}{5}$ and y-intercept $(0, −3)$

22. slope = 4 through the point $(1, 5)$

23. slope = $-\frac{3}{4}$ through the point $(−8, 12)$

24. through the points (–4, 3) and (6, –2)

25. through the points (2, –6) and (–3, 4)

Multiply each pair of binomials.

26. $(x-5)(x+2)$

27. $(x-9)(x-3)$

28. $(x-4)(x+8)$

29. $(3x-4)(2x+5)$

30. $(5x-1)(2x-7)$

Factor completely.

31. $x^2-9x+20$

32. $x^2+13x+40$

33. x^2-x-12

34. $5x^2+8x-21$

35. $8x^2+14x-15$

Chapter 2

Determine if the relation is a function.

36.

x	2	6	3	–1	2
y	5	–1	–3	8	7

37. $x^2-5y=20$

Evaluate each function.

38. For $f(x)=x^3+2x^2-3x+1$, find $f(-2)$.

39. For $g(x)=\sqrt{2x+3x^2}$, find $g(-2)$.

40. For $f(x)=\frac{3x-5}{x-2}$, find $f(0)$.

Given functions $f(x)=x+1$ and $g(x)=x^2-1$, find each of these new functions and give the domain of the new function.

41. $f+g(x)$

42. $g-f(x)$

43. $f\cdot g(x)$

44. $\frac{f}{g}(x)$

45. $\frac{g}{f}(x)$

Given $f(x)=\sqrt{4+x}$, $g(x)=x^2+4$, and $h(x)=x-4$, find each composition and give its domain.

46. $f\circ g(x)$

47. $g\circ f(x)$

48. $f\circ h(x)$

49. $h\circ h(x)$

50. $f(h(g(x)))$

Find the inverse of each function and give its domain.

51. $f(x)=8-3x$

52. $g(x)=\sqrt{x-4}-1$

53. $f(x)=(x-3)^2, x\geq 3$

54. $g(x)=\frac{2}{x-4}$

55. $f(x)=\frac{x+1}{x^2-1}$

Given the parent function $f(x) = |x|$, whose graph is shown, sketch a graph of each of the following.

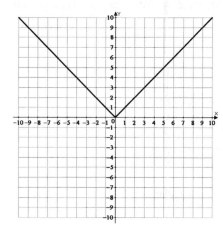

56. $g(x) = |x - 4| + 3$

57. $g(x) = 2 - |x + 1|$

58. $g(x) = 3|4 - x| + 2$

59. $g(x) = -\frac{2}{3}|x| + 7$

60. $g(x) = 5 - 3|x + 2|$

Chapter 3

Graph each inequality.

61. $y < -3x + 7$ 62. $y \geq \frac{5}{3}x - 2$ 63. $2x + 5y \leq 20$

64. $3x - 2y > 12$ 65. $-6y > 12 - 5x$

Find the line of best fit for each set of points. Use a graphing calculator or software if available.

66.

x	−1	−2	2	5	3	−4	−1	1	1
y	6	8	−3	−10	−6	12	8	1	2

67.

x	7	4	5	18	3	10	6	14	20
y	5.3	9.8	8.2	−10.1	11.7	0.7	7.6	−4.8	−11.3

68.

x	8	9	22	14	17	16	24	19	20
y	19	20	55	32	45	36	63	45	47

69.

x	38	30	27	31	39	28	32	26	39
y	80	85	88	83	82	86	81	84	79

70.

x	54	56	55	53	51	62	52	52	62
y	22	21	21.5	19.5	22.5	21.5	20.5	21	22.5

Graph each piecewise function.

71. $f(x) = \begin{cases} -4 & \text{if } x > 2 \\ 3 & \text{if } x \leq 2 \end{cases}$

72. $g(x) = \begin{cases} -\frac{2}{3}x + 4 & \text{if } x < 2 \\ \frac{1}{5}x - 2 & \text{if } x \geq 2 \end{cases}$

73. $y = \begin{cases} x - 5 & \text{if } x \geq 0 \\ 2x + 1 & \text{if } x < 0 \end{cases}$

74. $f(x) = \begin{cases} 1 - x & \text{if } x < -2 \\ x - 1 & \text{if } -2 \leq x \leq 4 \\ 4 & \text{if } x > 4 \end{cases}$

75. $g(x) = \begin{cases} -2x & \text{if } x \leq -3 \\ x & \text{if } -3 < x \leq -1 \\ 2 & \text{if } -1 < x \leq 1 \\ -x & \text{if } 1 < x \leq 3 \\ 2x & \text{if } x > 3 \end{cases}$

Chapter 4

Solve each equation.

76. $|8x + 5| = 19$

77. $|3 - 7x| + 9 = 13$

78. $|2x - 11| - 2 = 1$

79. $12 - \frac{1}{2}|x - 6| = -18$

80. $\frac{2}{3}|4x + 5| - 18 = 20$

Solve each inequality.

81. $|4 - x| \leq 12$

82. $|3x - 11| + 2 > 36$

83. $3 - 8|5t - 11| < -29$

84. $14 - \left|\frac{1}{2}x + 7\right| \geq 2$

85. $3|2 - 9x| + 2 < 50$

Use transformations to sketch the graph of each function.

86. $f(x) = \frac{1}{2}|x + 5|$

87. $g(x) = |x - 4| - 7$

88. $y = 1 - |x + 4|$

89. $f(x) = 2|x - 3| + 5$

90. $g(x) = 4 - \frac{1}{2}|x + 6|$

Chapter 5

Solve each system of equations by graphing.

91. $y = 5 - 2x$
    $4x + 5y = 7$

92. $3y = 2x - 12$
    $3x - 4y = 28$

93. $5x - 2y = -2$
    $x - 4y = 14$

94. $y + 6 = \frac{1}{2}x + 1$
    $x - y = 8$

95. $y = -x - 3$
    $4x + 7y = -6$

Solve each system of equations by substitution.

96. $y - \frac{1}{2}x = 6$
    $2x - 3y = -20$

97. $x + y = 11$
    $2x - 4y = -2$

98. $4x + 5y = 7$
    $y - 3x = 9$

99. $5x + y = 2$
    $3x - 2y = -17$

100. $5x - y = 15$
     $x - 7y = -31$

Solve each system by elimination.

101. $4x + 3y = 14$
     $9x - 2y = 14$

102. $2x - 5y = 16$
     $4x - 3y = 11$

103. $5x - 3y = -11$
     $7x + 6y = -12$

104. $7x - 6y = 13$
     $6x - 5y = 11$

105. $3x - 2y = 1$
     $-6x + 4y = -2$

Solve each system by substitution, elimination, or a combination of the two.

106. $x - y + 2z = -7$
     $x + y - z = 6$
     $x + 2y + z = 2$

107. $x + y + z = 6$
     $2x - y + z = 3$
     $x + 2y - 3z = -4$

108. $4x + y + 2z = 8$
     $6x - 2y + 4z = 2$
     $5x - 3y + 6z = 5$

109. $3x - 2y + z = 2$
     $3x - 2y + z = 2$
     $x + 4y - z = 4$

110. $3x + 5y - 2z = 10$
     $6x + 10y - 4z = 2$
     $x - 8y + z = 4$

Graph each system of inequalities. Mark the solution set clearly.

111. $y \leq 6 - x$
     $2x - 3y > 2$

112. $y < \frac{1}{2}x + 1$
     $x - 3 \leq y$

113. $x - 2y > 12$
     $3x + y < 15$

114. $y \leq 5x + 17$
     $7x + 33 \geq 6y$

115. $y > \frac{3}{4}x - 3$
     $2x - 5y \leq 1$

Determine the vertex of the feasible region that represents the maximum value of the objective function.

116. Profit = $12x + 16y$ subject to $12x + 4y \leq 50$, $4x + 10y \leq 60$, $x \geq 0$, and $y \geq 0$.

117. Profit = $11.50x + 6.25y$ subject to $8x + 2y \leq 50$, $10x + 5y \leq 80$, $x \geq 0$, and $y \geq 0$.

Determine the vertex of the feasible region that represents the minimum value of the objective function.

118. Cost = $4x + 5y$ subject to $5x + 10y \geq 60$, $8x + 2y \geq 40$, $0 \leq x \leq 10$, and $0 \leq y \leq 10$.

119. Cost = $70x + 60y$ subject to $10x + 4y \geq 50$, $12x + 5y \leq 80$, $y \leq x + 5$, $x \geq 0$, and $y \geq 0$.

120. A farmer has 100 acres of land on which to plant corn and soybeans. Each acre of corn will require 5 hours of labor, and each acre of soybeans will require 2 hours of labor. Each acre of corn will need to be irrigated with 2 gallons of water, and each acre of soybeans with 5 gallons. Seed costs $30 an acre for corn and $10 an acre for soybeans. There are 400 hours of labor and 400 gallons of water for irrigation available. The farmer has budgeted at least $500 to purchase seed. How many acres of corn and how many acres of soybeans should he plant to maximize profit if he makes a profit of $200 per acre of corn and $250 per acre of soy?

Chapter 6

Given the matrices $A = \begin{bmatrix} 3 & 1 \\ -2 & 0 \end{bmatrix}$, $B = \begin{bmatrix} 2 & -1 \\ 3 & -2 \\ 0 & 1 \end{bmatrix}$,

$C = \begin{bmatrix} 2 \\ -1 \end{bmatrix}$, $D = \begin{bmatrix} -1 & -3 & -2 \\ 0 & 2 & 1 \end{bmatrix}$, $E = \begin{bmatrix} 3 & -2 & -1 \end{bmatrix}$,

and $F = \begin{bmatrix} 2 & 1 \\ -3 & -1 \end{bmatrix}$, perform each calculation, if possible.

121. $[F] - [A]$

122. $[B] \cdot [D]$

123. $[A] \cdot [D]$

124. $[F] + [D]$

125. $[E] \cdot [B] \cdot [C]$

Find the determinant of each matrix, if possible.

126. $\begin{vmatrix} 2 & 1 \\ 3 & 4 \end{vmatrix}$

127. $\begin{vmatrix} -3 & 2 \\ 6 & -4 \end{vmatrix}$

128. $\begin{vmatrix} 8 & -3 \\ -2 & 5 \end{vmatrix}$

129. $\begin{vmatrix} 2 & -1 & 0 \\ 1 & 0 & -2 \\ 0 & 1 & 2 \end{vmatrix}$

130. $\begin{vmatrix} 1 & 3 & 7 \\ -2 & 6 & 4 \\ 3 & 7 & -1 \end{vmatrix}$

Solve by Cramer's rule.

131. $2x - 3y = 3$
 $4x - 2y = 10$

132. $3x + y = -2$
 $-3x + 2y = -4$

133. $4x - 7y = 3$
 $5x + 2y = -3$

134. $2x - 3y + z = 1$
 $3x - y - z = 4$
 $4x - 6y + 2z = 3$

135. $2x - y + 3z = 4$
 $x - 5y - 2z = 1$
 $-4x - 2y + z = 3$

Find the solution of the system by putting the augmented matrix in reduced row echelon form.

136. $5x - 2y = 4$
 $-10x + 4y = 1$

137. $3x - 4y = 7$
 $6x - 2y = 5$

138. $2x - 3y = 4$
 $4x - 5y = 3$

139. $x + 2y + z = 2$
 $x + y - z = 6$
 $x - y + z = -4$

140. $x + y = 6$
 $2x + z = 3$
 $2y - 3z = -4$

Find the inverse of each matrix, if possible.

141. $\begin{bmatrix} 2 & 1 \\ 4 & 3 \end{bmatrix}$

142. $\begin{bmatrix} 6 & 1 \\ 2 & \frac{1}{2} \end{bmatrix}$

143. $\begin{bmatrix} 4 & 4 \\ 1 & 1 \end{bmatrix}$

144. $\begin{bmatrix} 2 & 0 & -1 \\ 3 & 1 & 2 \\ 5 & -2 & 1 \end{bmatrix}$

145. $\begin{bmatrix} 1 & 3 & 7 \\ -2 & 6 & 4 \\ 3 & 7 & -1 \end{bmatrix}$

Solve each system by using an inverse matrix.

146. $3x + y = 2$
$2x + y = 0$

147. $3x - 7y = 2$
$-4x + 6y = -6$

148. $5x - 2y = 7$
$3x + y = 2$

149. $5x - 3y - 6z = 5$
$4x - 6y - 3z = 4$
$-x + 9y + 9z = 7$

150. $4x - 6y + 8z = 4$
$5x + y - 2z = 4$
$6x - 8y + 12z = 6$

Chapter 7

Evaluate each expression.

151. $27^{\frac{2}{3}}$

152. $(16)^{\frac{3}{4}}$

153. $\left(\frac{81}{49}\right)^{-\frac{1}{2}}$

154. $\left(\frac{8}{1,000}\right)^{-\frac{5}{3}}$

155. $(-32)^{\frac{4}{5}}$

Simplify each expression.

156. $\sqrt{81a^4b^8c^7}$

157. $\sqrt[3]{81x^3y^9z^4}$

158. $\left(\frac{16a^5b^4}{27z^9}\right)^{\frac{1}{3}}$

159. $\sqrt[5]{\frac{128x^3y^7}{z^{12}}}$

160. $\left(\frac{9x^7y^3}{125z^4}\right)^{-\frac{3}{2}}$

Rationalize each denominator and leave the expression in simplest form.

161. $\frac{\sqrt{x}}{\sqrt{x}+3}$

162. $\frac{\sqrt{2}+7}{\sqrt{2}-7}$

163. $\frac{2\sqrt{3}-\sqrt{7}}{3\sqrt{3}+\sqrt{7}}$

164. $\frac{3}{\sqrt[3]{4}}$

165. $\frac{1+\sqrt[3]{4}}{\sqrt[3]{4}}$

Solve each equation.

166. $\sqrt{4x+5}+2=7$

167. $t - 6 = \sqrt{t-4}$

168. $\sqrt{x+2} = \sqrt{x+3}-1$

169. $\sqrt[3]{4x+5} = 3$

170. $\sqrt[4]{6x+7} - \sqrt[4]{x+2} = 0$

Sketch a graph of each function.

171. $f(x) = \sqrt{x+2} - 4$

172. $y = -2\sqrt{x-5}$

173. $y = 9 - \sqrt{3-x}$

174. $f(x) = 4\sqrt[3]{x+1}$

175. $g(x) = 3 - \frac{1}{2}\sqrt[3]{x-5}$

Chapter 8

Solve by the square root method.

176. $4x^2 - 64 = 0$

177. $(2x - 3)^2 = 25$

178. $(3x + 2)^2 = 48$

179. $x^2 - 6x + 9 = 25$

180. $(x - 7)^2 + 5 = 86$

Solve by completing the square.

181. $x^2 + 6x = 3$

182. $x^2 + 3x - 4 = 0$

183. $5x^2 - 3x - 2 = 0$

184. $x^2 + 5x - 2 = 0$

185. $3x^2 - 10x + 7 = 0$

Solve by the quadratic formula.

186. $6x^2 + 7x + 2 = 0$

187. $\frac{1}{2}x^2 + x = \frac{1}{3}$

188. $6z^2 + 7z - 5 = 0$

189. $0.01x^2 + 0.06x - 0.08 = 0$

190. $2x - 3 = 3x^2$

Determine the number and type of solutions, but do not solve.

191. $x^2 - 5x - 3 = 0$

192. $3x^2 - 8x + 1 = 0$

193. $4x^2 - 3x + 5 = 0$

194. $5x^2 + 12x - 1 = 0$

195. $7x^2 - 5x + 2 = 0$

Perform the indicated operations and leave the expression in simplest form.

196. $(7 + 5i) - (5 - 3i) + (1 + 4i)$

197. $(3 + 2i)(2 - i)$

198. $\frac{-2i}{5-3i}$

Solve each equation, putting complex solutions in simplest form.

199. $x^2 - 2x + 5 = 0$

200. $2x^2 + 3x + 3 = 0$

Sketch a graph of each function.

201. $f(x) = 2(x + 1)^2$

202. $y = -\frac{1}{2}(x-3)^2 + 2$

203. $g(x) = 3(x - 4)^2 + 1$

204. Complete the square to put $y = x^2 - 4x + 7$ in vertex form and sketch the graph.

205. Complete the square to put $f(x) = -2x^2 + 12x - 10$ in vertex form and sketch the graph.

Find the equation of a quadratic function that passes through the given points.

206. Vertex (2, 6) and passing through (5, –3)

207. Vertex (–5, 4) and passing through (–1, 12)

208. Vertex (1, –3) and y-intercept of –5

209. Passing through the points (–2, 17), (1, –1), and (6, 9)

210. Passing through the points (–2, –18), (–1, –2), and (2, 10)

Chapter 9

Put each equation in vertex form.

211. $x^2 + 6x - 3y = 3$

212. $5y - x^2 = 4x - 16$

213. $5y^2 + 20y + 50 = 10x$

214. $2x^2 + y - 34x + 99 = 0$

215. $2y^2 + 4y - x - 9 = 0$

Sketch the graph of each parabola. Show the focus and directrix.

216. $y = -2(x - 4)^2 - 3$

217. $x + 1 = 2(y - 3)^2$

218. $y = -\frac{1}{8}(x - 1)^2 + 4$

219. $x - 4 = \frac{1}{2}(y + 5)^2$

220. $x + 2 = -(y + 6)^2$

Sketch the graph of each circle.

221. $(x - 8)^2 + (y + 8)^2 = 64$

222. $4(x + 1)^2 + 4(y + 5)^2 = 100$

223. $(x - 7)^2 + (y + 1)^2 = 36$

224. $x^2 + y^2 - 6y - 40 = 0$

225. $x^2 + y^2 - 6x = 0$

Put the equation of each ellipse in standard form.

226. $4x^2 + 25y^2 = 100$

227. $16x^2 - 96x + y^2 + 128 = 0$

228. $9x^2 + 4y^2 + 8y + 4 = 36$

229. $9x^2 - 90x + 36y^2 + 144y + 45 = 0$

230. $7x^2 + 4y^2 + 98x + 64y + 571 = 0$

Sketch the graph of each ellipse. Mark the foci.

231. $\frac{(x+1)^2}{4} + \frac{(y-5)^2}{9} = 1$

232. $\frac{(x-2)^2}{25} + \frac{(y-3)^2}{16} = 1$

233. $\frac{(x+7)^2}{36} + \frac{(y+9)^2}{100} = 1$

234. $\frac{(x-6)^2}{49} + \frac{(y+9)^2}{64} = 1$

235. Sketch a graph and find an equation for the ellipse with a center at (2, 1), focus at (2, 5), and vertex at (2, 6).

Put the equation of each hyperbola in standard form.

236. $x^2 - y^2 + 2x - 2y - 3 = 1$

237. $4x^2 - y^2 + 24x - 6y + 23 = 0$

238. $y^2 + 10y - 4x^2 + 56x = 187$

239. $9x^2 - 16y^2 + 36x - 96y - 252 = 0$

240. $25y^2 - 4x^2 + 100y + 56x - 196 = 0$

Sketch the graph of each hyperbola. Show the asymptotes and mark the foci.

241. $\frac{x^2}{25} - \frac{y^2}{64} = 1$

242. $\frac{(y+1)^2}{16} - \frac{(x-5)^2}{9} = 1$

243. $\frac{(x+3)^2}{49} - \frac{y^2}{4} = 1$

244. $81(y - 6)^2 - 4(x + 11)^2 = 324$

245. Sketch a graph and find an equation for the hyperbola that has vertices at (±6, 2), if its asymptotes are the lines $y = \pm\frac{5}{4}x + 2$.

Solve each system.

246. $4x^2 + 9y^2 = 36$
     $y = -\frac{1}{2}x + 2$

247. $x^2 + y^2 = 16$
     $9x^2 - 4y^2 = 36$

248. $x^2 + y^2 = 4$
     $x - 2y = 4$

249. $y = x^2 - 4$
     $x^2 + y^2 = 4$

250. $x + y = 2$
     $x^2 - y^2 = 4$

Chapter 10

Factor each expression.

251. $64x^3 + 1$

252. $8t^3 - 27$

253. $8x^3 + 12x^2 + 6x + 1$

254. $27y^3 + 125$

255. $x^3 - 343$

List the possible rational zeros of each polynomial.

256. $f(x) = 4x^3 - 6x^2 + 5x - 3$

257. $y = x^3 + 3x^2 - 4x - 12$

258. $g(x) = 2x^3 + 3x^2 - 8x - 12$

259. $y = x^3 + 5x^2 - 4x - 20$

260. $f(x) = 3x^3 + 2x^2 - 27x - 18$

Determine the maximum number of positive real zeros and the maximum number of negative real zeros for each polynomial.

261. $f(x) = x^3 + 2x^2 - 25x - 50$

262. $g(x) = 4x^3 + 12x^2 - 9x - 27$

263. $f(x) = x^3 + 4x^2 - 9x - 36$

264. $g(x) = 9x^3 + 18x^2 - 4x - 8$

265. $f(x) = x^3 + 2x^2 + 9x + 18$

Divide using long division. Is the divisor a factor of the dividend?

266. $(x^2 + 4x - 8) \div (x - 3)$

267. $(6x^2 + 7x - 18) \div (3x - 4)$

268. $(3x^3 - 5x^2 + 2x - 1) \div (x - 2)$

269. $(2x^3 - 9x^2 + 11x - 6) \div (2x^2 - 3x + 2)$

270. $(a^4 + a^3 - 1) \div (a + 2)$

Use synthetic division to determine if the given value of x is a zero of the polynomial.

271. $f(x) = x^2 - 5x - 6$ given $x = -2$

272. $f(x) = 3x^2 - 4x + 1$ given $x = 1$

273. $f(x) = x^3 + 2x^2 + 3x + 4$ given $x = 2$

274. $f(x) = 3x^3 - x^2 + 2x + 5$ given $x = 3$

275. $f(x) = 2x^3 + x - 3$ given $x = 1$

Find all zeros of the polynomial function. One zero is given.

276. $f(x) = 9x^4 + 4x^3 - 3x^2 + 2x$ given $x = -1$

277. $g(x) = 15x^4 - 25x^3 + 10x^2$ given $x = 1$

278. $f(x) = 3x^3 - 13x^2 + 8x + 12$ given $x = 2$

279. $g(x) = 3x^3 - 18x^2 + 33x - 18$ given $x = 3$

280. $f(x) = 2x^3 + 12x^2 + 22x + 12$ given $x = -2$

Find the solution set for each polynomial inequality.

281. $2x^3 + 17x^2 + 41x + 30 \le 0$

282. $2x^3 + 17x^2 + 41x + 30 > 0$

283. $20x^4 + 13x^2 - 15 \le 0$

284. $x^3 + 5x^2 - 4x - 20 \ge 0$

285. $45x^4 \le 30x^3 - 5x^2$

Sketch the graph of each function.

286. $y = 2x^3 + 20x^2 + 50x$

287. $y = x^3 + 5x^2 - 9x - 45$

288. $y = x^3 - 17x^2 + 91x - 147$

289. $y = 3x^3 + 2x^2 - 27x - 18$

290. $y = x^3 + 5x^2 - 4x - 20$

Chapter 11

Give the domain of each rational expression and then simplify.

291. $\frac{t-5}{5-t}$

292. $\frac{x^2-25}{5-x}$

293. $\frac{y^2-y-6}{y^2-4}$

294. $\frac{x^2-16}{6x+24}$

295. $\frac{x^2-4x-12}{x^2+8x+12}$

Perform addition or subtraction as indicated and leave the answers in simplest form.

296. $\frac{-2}{x^2-2x-3} + \frac{3}{x^2-9}$

297. $\frac{x-4}{2x-6} + \frac{3}{x^2-9}$

298. $\frac{2x-4}{x^2+5x+4} - \frac{x-4}{x^2+6x+8}$

299. $\frac{2x-2}{x^2+4x+3} - \frac{x-1}{x^2+5x+6}$

300. $2 - \frac{9}{3x+1}$

Multiply and leave the answers in simplest form.

301. $\frac{x+5}{x^2-25} \cdot \frac{x-5}{x^2-10x+25}$

302. $\frac{3y^2-3y}{3y-12} \cdot \frac{y^2-2y-8}{y^2+3y+2}$

303. $\frac{2x^2-4x}{2x^2-2} \cdot \frac{x^2-2x-3}{x^2-5x+6}$

304. $\frac{3x-12}{x^2-4} \cdot \frac{x^2+6x+8}{x-4}$

305. $\frac{y-1}{y^2-y-6} \cdot \frac{y^2+5y+6}{y^2-1}$

Divide and leave the answers in simplest form.

306. $\frac{a^2-5a+6}{a^2-2a-3} \div \frac{a-5}{a^2+3a+2}$

307. $\frac{2x^2-5x-12}{4x^2+8x+3} \div \frac{x^2-16}{2x^2+7x+3}$

308. $\frac{9t^2-1}{6t^2+7t-3} \div \frac{27t^3+1}{8t^3+27}$

309. $\frac{x^5-x^2}{5x^5-5x} \div \frac{10x^4-10x^2}{2x^4+2x^3+2x^2}$

310. $\frac{4y^2-y^3}{16-y^2} \div \frac{64+y^3}{16+8y+y^2}$

Simplify each complex fraction as completely as possible.

311. $\frac{\frac{1}{x}+\frac{2}{y}}{\frac{2}{x}+\frac{1}{y}}$

312. $\frac{\frac{x-5}{x^2-4}}{\frac{x^2-25}{x+2}}$

313. $\frac{1-\frac{1}{x}}{\frac{1}{x}}$

314. $\frac{1+\frac{1}{x-2}}{1-\frac{3}{x+2}}$

315. $\frac{\frac{2a}{3a^3-3}}{\frac{4a}{6a-6}}$

Solve by cross-multiplying.

316. $\frac{2}{a+5} = \frac{1}{3}$

317. $\frac{x+1}{x+2} = \frac{x-3}{x-4}$

318. $\frac{-2}{x-3} = \frac{x+3}{4}$

319. $\frac{3}{a-2} = \frac{2}{a-5}$

320. $\frac{y-2}{y-5} = \frac{2}{y+5}$

Solve each equation. Be sure to check for extraneous solutions.

321. $\frac{x}{x+1} - \frac{1}{2} = \frac{-1}{x+1}$

322. $\frac{x}{x^2-9} - \frac{1}{x+3} = \frac{1}{4x-12}$

323. $\frac{5}{x^2-3x+2} - \frac{1}{x-2} = \frac{1}{3x-3}$

324. $\frac{2}{x^2-7x+12} - \frac{1}{x^2-9} = \frac{4}{x^2-x-12}$

325. $\frac{3}{y-4} - \frac{2}{y+1} = \frac{5}{y^2-3y-4}$

Give the domain of each function. Find all the asymptotes: vertical, horizontal, and oblique (if they exist).

326. $f(x) = \frac{-1}{x+2} - 3$

327. $g(x) = \frac{x-3}{x^2-1}$

328. $f(x) = \frac{-2x}{x+5}$

329. $g(x) = \frac{x^2-x-6}{x+1}$

330. $f(x) = \frac{x^2-9}{x-3}$

Sketch a graph of each function. Show asymptotes and intercepts.

331. $f(x) = \frac{4}{x+2} - 3$

332. $g(x) = \frac{2x+1}{x-3}$

333. $f(x) = \frac{x+5}{x-3}$

334. $g(x) = \frac{x^2-4}{x-3}$

335. $f(x) = \frac{3x}{x^2-x-12}$

Chapter 12

Rewrite in logarithmic form.

336. $5^4 = 625$

337. $9^y = 6,561$

Rewrite in exponential form.

338. $\log_2 1{,}024 = 10$

339. $\log_3 x = 2$

340. Use your calculator and the change-of-base rule to find $\log_7 325$.

Expand each logarithm to a sum, difference, and/or multiple of logs.

341. $\log_2\left(\frac{x^4}{y^3\sqrt{z}}\right)$

342. $\log\left(\frac{x^4\cdot\sqrt[3]{y}}{z^3}\right)$

343. $\log_8\sqrt[3]{xy^5}$

344. $\log_3\left(\frac{x^2 y}{\sqrt{z}}\right)$

345. $\log\left(\frac{x\sqrt{y}}{z^4}\right)$

Condense each expression to a single log.

346. $3\log_4 x + \log_4 y - 2\log_4 z$

347. $2\log a + 3\log b - \frac{1}{3}\log c$

348. $\frac{1}{2}\log(x) + \log(y) - 4\log(z)$

349. $3\ln(x) + \ln(x + 5) - 2\ln(x + 4)$

350. $2\log_2(x - 1) + \log_2(x + 5) - \log_2(x + 3)$

Solve by writing both sides as powers of the same base.

351. $8^{x+1} = 4$

352. $9^{2x-4} = 27^3$

Solve by taking the log of both sides.

353. $12^{-x} = 5$

354. $3^{2x+1} = 2$

355. $10^{3x-4} = 15^{x-2}$

Each of these equations is quadratic in form. Solve each equation for x.

356. $2^{2x} - 3(2^x) + 2 = 0$

357. $3(5^{2x}) + 14(5^x) - 5 = 0$

358. $2(3)^{2x} = (3^{x+1}) + 20$

359. $e^{2x} - 11(e^x) = -30$

360. $3(2^{2x}) = -2^{x+1} + 2^3$

Solve each logarithmic equation.

361. $\log_2(x + 2) + \log_2 x = 3$

362. $\log_2(x + 3) + \log_2 x = 2$

363. $\log_3(x + 3) - \log_3(x - 1) = 1$

364. $\log_8(x) + \log_8(x - 3) = \frac{2}{3}$

365. $\log_4(x - 2) - \log_4(x + 1) = 1$

Sketch a graph of each function. Draw as much of the graph as you can before using a graphing calculator or software.

366. $f(x) = 2^{x-3} + 1$

367. $g(x) = 4^{-x} - 3$

368. $f(x) = 3\log_2(x - 5)$

369. $g(x) = \log_3(2 - x) + 1$

370. $f(x) = -2\ln(x + 3) - 4$

Use the given formula to solve each problem.

371. According to the College Board, the average tuition in U.S. private colleges for the 2014–2015 school year was \$31,231. In order to have that amount of money available on a student's 18th birthday, how much would have to be invested at the student's birth, at a rate of 6% compounded continuously? (Use $A = Pe^{rt}$.)

372. How long will it take to double your money if it is invested at an annual interest rate of 4% compounded monthly? (Use $A = P\left(1+\frac{r}{n}\right)^{nt}$.)

373. On initial inspection, a colony of bacteria had 300 bacteria present. When examined 2 days later, the population of the colony had tripled. If reproduction continues at the same rate, find the rate and predict the population after 7 days. (Use $A = Pe^{rt}$.)

374. Carbon 14 has a half-life of 5,730 years. Use $A = Pe^{rt}$ to find the rate of decay, and then determine the approximate age of a fossil that has 74% of its original Carbon 14 remaining.

375. The Richter scale is a method of rating the power of an earthquake based on the amount of energy released. The formula $M = \log\left(\frac{I}{I_0}\right)$ gives the magnitude M, as a function of the intensity of energy released, I, and the baseline intensity, $I_0 = 10^{-4}$. The famous San Francisco earthquake of 1906 is estimated to have measured 7.8 on the Richter scale. In 1989, another earthquake occurred near San Francisco and was seen on national television during World Series broadcast. The 1989 quake measured 6.9 on the Richter scale. How many times more energy was released in the 1906 quake as compared to the 1989 quake?

Chapter 13

Use the basic counting principle to calculate the number of different outcomes in each situation.

376. Area codes are three-digit codes that may not begin with 0. How many different area codes are possible?

377. If, in addition to the rule that the first digit of an area code cannot be 0, a rule is added saying that the final digit must be even, how many area codes are possible?

378. If a state requires that license plates begin and end with a letter and contain four digits in between, how many license plates are possible?

379. A company that requires its employees to wear a uniform provides options for its female employees. Women may choose to wear a skirt or pants, and may choose from a solid-color shirt, a patterned blouse, or a T-shirt. In addition, women may choose a blazer, a sweater, or no additional layer. How many uniforms are possible?

380. A game show requires a contestant to guess the price of a prize by choosing each digit in the price from a group of digits presented. Suppose the price contains four digits, with no repeats. If the contestant is given the correct four digits and has to determine the order, how many possibilities are there to choose from? If the contestant is instead presented with the four correct digits and two additional digits that are not in the price, how many possibilities exist?

Find the number of permutations or combinations as indicated.

381. $_4P_2$

382. In how many different orders can the seven members of the cast of a play appear for their curtain calls?

383. $_5C_2$

384. $_8C_8$

385. The 22 employees at a small business want to choose a committee of 5 to negotiate for better working conditions. How many different committees of 22 are possible?

Use Pascal's triangle to find the number of combinations.

386. $_4C_3$

387. $_5C_3$

388. $_6C_2$

389. $_7C_3$

390. $_9C_6$

Find the indicated probability.

391. A card is drawn from a standard deck of 52 playing cards, recorded, and replaced, and then a second card is drawn. What is the probability that both cards are aces?

392. If the experiment described in the preceding question is repeated without replacing the first card, what is the probability that both cards are aces?

A group of people were asked about their ability to speak English and Spanish. Their responses are summarized in the following table, divided by the gender of the respondent. Use the information in the table to answer questions 393 through 395.

	English Only	Spanish Only	Both English and Spanish	Total
Men	210	200	90	500
Women	315	150	35	500
Total	525	350	125	1,000

393. If a person is chosen at random from this group, what is the probability that he or she speaks both English and Spanish?

394. What is the probability that a person chosen at random is a woman who speaks only Spanish?

395. What is the probability that a person chosen at random speaks only English, given that the person is a man?

Use the binomial theorem and Pascal's triangle to expand each power of a binomial.

396. $(x-2)^3$ 397. $(2x+1)^3$ 398. $(2x+3)^4$

399. $(x+2)^5$ 400. $(2x-1)^6$

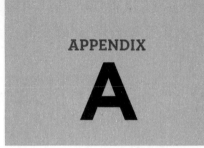
Glossary

absolute value   The distance of a number from 0, regardless of direction. The size or magnitude of a number without regard to its sign. Formally, the absolute value function is defined as a piecewise function:

$$|x| = \begin{cases} x & \text{if } x \geq 0 \\ -x & \text{if } x < 0 \end{cases}$$

addend   Each of the numbers that are being added.

algorithm   A list of steps necessary to perform a process.

associative property   A property of addition or multiplication that says when adding or multiplying more than two numbers, you may group them in different ways without changing the result.

asymptote   A line that a graph approaches very closely but does not cross.

augmented matrix   A rectangular matrix with R rows and C columns formed by appending an additional column to a square matrix.

axis   A vertical or horizontal line that divides the coordinate plane into sections. The horizontal line is called the x-axis and the vertical line is called the y-axis.

axis of symmetry   An imaginary line passing through the vertex of a parabola, dividing the parabola into two sections that are reflections of one another.

base   When an exponent is used to show repeated multiplication, the number to be multiplied is the base of the power.

basic counting principle   The number of different ways a series of choices can be made is the product of the number of different options for each choice.

binary operation   A process that works on two numbers at a time.

binary system   A place value system based on the number 2.

binomial theorem   The pattern of terms of the polynomial produced when a binomial is raised to a power:

$$(a+b)^n = a^n + \binom{n}{1}a^{n-1}b + \binom{n}{2}a^{n-2}b^2 + \ldots + \binom{n}{n-2}a^2b^{n-2} + \binom{n}{n-1}ab^{n-1} + \binom{n}{n}b^n$$

canceling   The process of simplifying a multiplication of fractions by dividing a numerator and a denominator by a common factor.

Cartesian plane   A system that identifies every point in the plane by an ordered pair of numbers. Also called the *coordinate plane*.

change-of-base rule   The log base b of a number N can be written in terms of another base, c, by dividing the log base c of N by the log base c of b. You change the base by dividing the log of the number by the log of the base.

circle   The set of all points at a fixed distance called the *radius* from a center point.

closure   A property that says adding or multiplying two numbers of a set gives you another number in the set.

coefficient   The numerical part of a term. Technically, a term is made up of a numerical coefficient and a variable coefficient, but usually the variable coefficient is just called the *variable* and the numerical coefficient is called the *coefficient*.

combination   A collection of objects for which order is not significant.

common denominator   A multiple of the denominators of two or more fractions.

common fraction   Fractions written as a quotient of two integers.

commutative property   A property of addition or multiplication that says reversing the order of the two numbers will not change the result.

completing the square   A method of solving quadratic equations by adjusting the constant term of the quadratic polynomial so it becomes a perfect square trinomial. It can then be written as the square of a binomial and solved by the square root method.

complex fraction   A fraction that contains one or more fractions in its numerator and/or denominator.

complex number   A number of the form $a + bi$, where a is a real number and bi is an imaginary number.

composite function A new function formed by two functions working in sequence.

compound inequality Two or more inequalities connected by a conjunction, such as *and* or *or.*

conditional probability The probability the event occurs, given that another event has already occurred.

conic section The intersection of a plane and a cone. The intersection forms a figure that can be described by a quadratic relation.

conjugate A sum and difference of two terms, involving a radical or imaginary number, whose product contains no radical or imaginary numbers.

conjunction A compound inequality in which the word *and* is used.

constant A term containing only numbers and no variables.

constraint Limitations on the activity, represented by an inequality.

coordinate system A system that locates every point in the plane by an ordered pair of numbers, (x, y). Also called the *Cartesian plane.*

counting numbers Also known as the natural numbers, the set of numbers that include 1, 2, 3, and all larger whole numbers. They are the numbers you use to count.

Cramer's rule A method of solving a system of equations using the determinants of matrices derived from the system.

cross-multiplying Finding the product of the means and the product of the extremes of a proportion, and saying those products are equal.

decimal fraction Fractions written in the base-10 system with digits to the right of the decimal point.

decimal system A place value system in which each position a digit can be placed is worth 10 times as much as the place to its right.

degree For a monomial with one variable, the power of the variable.

denominator The number below the division bar in a fraction that tells how many parts the whole was broken into, or what kind of fraction you have.

determinant A single number associated with a square matrix, found by operations with the numbers on the diagonals.

difference The result of a subtraction problem.

digit A single symbol that tells how many.

dimension When referring to a matrix, the number of rows by the number of columns.

discontinuity   A break in the graph of a function that occurs at a value excluded from the domain.

discriminant   The radicand of the quadratic formula, $b^2 - 4ac$, whose sign tells whether the quadratic equation has two real solutions ($b^2 - 4ac > 0$), one real solution ($b^2 - 4ac = 0$), or two nonreal solutions ($b^2 - 4ac < 0$).

disjunction   A compound inequality in which the two inequalities are joined by *or*.

distributive property   The product of a number and a sum (or difference) can be found by multiplying the number by each term of the sum (or difference). In symbols, this is $a(b + c) = ab + ac$.

dividend   In a division problem, the polynomial that is divided by the divisor.

divisor   In a division problem, the number you divide by.

domain   The set of all values that can be substituted in a function for x—that is, all possible inputs.

domain, rational expression   The set of all real numbers that can be substituted in a rational expression for the variable without making the denominator equal to 0.

double root   A solution that occurs twice for the same equation. This is also known as a solution with a multiplicity of 2.

elimination method   A method of solving a system that adds the two equations, or multiples of the equations, so the resulting equation contains only one variable. The value of that variable is found and substituted back in to find the other variable.

ellipse   An oval shape formed by all the points for which the sum of the distances from two focal points is constant.

equation   A mathematical sentence, which often contains a variable.

exponent   The small raised (superscript) number that tells how many times to use the base number as a factor. *See also* power.

exponential equation   An equation containing one or more terms with a variable in the exponent position.

exponentiation   Using each side of an equation as an exponent on the same base, creating a new, equivalent equation of equal powers.

expression   Any mathematical calculation.

extended ratio   Several related ratios condensed into one statement. The ratios $a{:}b$, $b{:}c$, and $a{:}c$ make the extended ratio $a{:}b{:}c$.

extraneous solution   A solution, which is produced by legitimate algebraic techniques, that does not satisfy the original equation.

extremes   The first and last numbers of a proportion.

factor   Each of the numbers or variables being multiplied.

factor tree   A method of finding the prime factorization of a number by starting with one factor pair and then factoring each of those factors, continuing until no possible factoring remains.

factorial   The product of positive integers in sequence. If n is a positive integer, n factorial (or $n!$) is the product of positive whole numbers from 1 to n.

factoring   The process of rewriting an integer as the product of two integers, or a polynomial as the product of two polynomials of a lesser degree.

feasible region   The area that is shaded by all the inequalities that represent the constraints. Each point is a feasible, or workable, solution within the constraints.

FOIL   An acronym for First, Outer, Inner, Last. It summarizes the four multiplications necessary when multiplying two binomials.

fraction   A symbol that represents part of a whole.

function   A relation in which each number in the domain has only one partner from the range.

function, linear   A function of the form $f(x) = mx + b$. It defines the value of the output variable as a multiple of the input variable, possibly plus or minus a constant.

greatest common factor   The largest number that is a factor of two numbers.

horizontal asymptote   A horizontal line that the ends of the graph of a rational function will approach very closely, either from above or below.

hyperbola   A pair of curves, reflections of one another, which curve away from one another around two focal points. The distances between a point of the hyperbola and each of the two focal points subtract to a constant value.

identity   A property that says adding 0 or multiplying by 1 leaves a number unchanged.

identity matrix   A square matrix with ones on the major diagonals and zeros elsewhere. Multiplying a matrix and an identity matrix of appropriate size leaves the matrix unchanged.

imaginary numbers   Multiples of the imaginary unit, i.

imaginary unit i; the symbol for the square root of -1.

improper fraction   A fraction whose value is more than 1. The numerator is larger than the denominator.

index A small number appearing in the crook of the radical sign that tells what power is being undone. If no index appears, the radical indicates a square root.

inequality A sentence that compares two expressions that are not equal and shows which one is larger.

inequality, compound Two inequalities connected by a conjunction, such as *and* or *or*.

inequality, linear A statement that the relationship between two expressions made up of constants and first-degree variable terms is unequal, with one expression larger than the other.

input The numbers from the domain that a function works on.

integers The set of numbers that includes the positive whole numbers, the negative whole numbers, and 0.

interest Money you pay for the use of money you borrow, or money you receive because you've put your money into a bank account or other investment.

interest rate The percent of the principal that will be paid in interest each year.

inverse functions Two functions whose composition, in either order, equals x.

inverse matrix Two square matrices whose product, in either order, is an identity matrix.

inverse operation An operation that reverses the work of another.

irrational numbers Any numbers that cannot be expressed as the ratio of two integers.

latus rectum A line segment parallel to the directrix of a parabola, which has the focus as its midpoint and whose endpoints rest on the parabola.

least common denominator The least common multiple of two or more denominators.

least common multiple The smallest number that has each of two or more numbers as a factor.

like terms Terms that have the same variable raised to the same power.

line A set of points that has length but no width or height.

line of best fit The line for which the total of the squares of the residuals is a minimum.

linear equation in one variable A mathematical sentence that says two expressions made up of constants and first-degree variable terms are equal.

linear function A function of the form $f(x) = mx + b$. It defines the value of the output variable as a multiple of the input variable, possibly plus or minus a constant.

linear inequality A statement that the relationship between two expressions made up of constants and first-degree variable terms is unequal, with one expression larger than the other.

linear inequality in two variables   A sentence that describes a relationship between two variables in which one variable is greater than or less than an expression involving the other variable.

linear programming   A technique for finding the best, or optimal, solution in a situation where choices are limited by a set of constraints but many solutions are possible within those constraints.

linear system   A set of two or more linear equations, each involving the same variables. The solution of the system is a set of values that satisfy all the equations in the system.

logarithm   The exponent that should be placed on a specified base to generate a particular number. L is the $\log_b N$ if $N = b^L$.

long division   A technique that achieves the division through a series of estimated quotients and repeated subtraction. The dividend polynomial is divided by the divisor to produce a quotient and possibly a remainder.

matrix   A device that displays numbers organized into rows and columns.

mean   The arithmetic average of a group of numbers, found by adding all the numbers and dividing by the number of numbers in the group.

means of a proportion   The two middle numbers in a proportion.

minuend   The first number in a subtraction problem.

mixed number   A whole number and a fraction, written side by side, representing the whole number plus the fraction.

monomial   A constant, a variable, or a product of constants and variables.

multiplicative inverse   When a fraction or expression is inverted or flipped, exchanging the numerator and denominator. Two rational numbers are multiplicative inverses if their product is 1. Two rational expressions are multiplicative inverses if they multiply to 1 over a domain on which both expressions are defined.

multiplicative inverse of a square matrix   Another square matrix such that $[A] \cdot [B] = [B] \cdot [A]$ = an identity matrix.

multiplicity   The number of times a solution occurs. It corresponds to the power to which the corresponding factor is raised.

natural numbers   The set of numbers that include 1, 2, 3, and all larger whole numbers. They are the numbers you use to count, which is why the natural numbers are also called the *counting numbers*.

negative exponent   On a nonzero base, represents a fraction with a numerator of 1 and a denominator of the base raised to the corresponding positive power.

nonlinear system   A system of equations that contains at least one equation of degree 2 or higher.

number line   A line divided into segments of equal length and labeled with numbers, usually the integers. Positive numbers increase to the right of 0, and negative numbers decrease to the left of 0.

numerator   The number above the bar in a fraction that tells you how many of that denomination are present.

objective function   An equation that defines what you want to maximize or minimize in choosing the best solution.

optimal solution   The point within the feasible region that maximizes or minimizes the objective function.

order of operations   An agreement that the order for simplifying or solving a problem is expressions in parentheses, exponents, multiplication and division as they appear from left to right, and finally addition and subtraction from left to right.

ordered pair   Two numbers, usually designated as x and y, that locate a point in a coordinate system.

origin   The intersection point of the four quadrants of a plane.

output   The numbers in the range that the function produces.

parabola   A conic section described as the set of points equidistant from a point called the *focus* and a line called the *directrix*.

parallel lines   Lines on the same plane that never intersect.

parent graph   The simplest graph that meets the description of a family of functions.

partial product   In a multiplication of polynomials, or in the multiplication of multi-digit numbers, the polynomial produced by multiplying by just one digit or term.

Pascal's triangle   A triangular arrangement of numbers that can be generated by starting and ending each row with a 1 and filling the row with sums of pairs of elements in the rows above. The triangle contains the number of combinations of n things taken 0, 1, …, n at a time, which are the coefficients necessary to form powers of binomials.

PEMDAS   A mnemonic or memory device to help you remember the order of operations: parentheses, exponents, multiplication and division, addition and subtraction.

percent   A ratio that compares numbers to 100. For instance, 42% means 42 out of 100, or 42:100.

permutation   An order or arrangement. The digits 2, 5, and 8 can be arranged in six different orders—258, 285, 528, 582, 825, and 852—meaning there are six permutations of those three digits.

perpendicular lines   Lines that meet to form a right angle.

Pert formula   A model for continuous growth or decay, based on the original amount, P; the rate of growth or decay, r; and the time, t. For growth, the rate is positive; for decay, it is negative.

piecewise function   A function which produces outputs according to different rules depending on what section of the domain the input comes from. The individual rules may be linear or nonlinear functions, or constant functions.

place value system   A number system in which the value of a symbol depends on where it is placed in a string of symbols.

plane   A flat surface that has length and width but no thickness.

point   A position in space that has no length, width, or height.

polynomial   A sum of terms, each of which is a product of a real number and a non-negative integer power of a variable.

power   A way to tell how many times a number should be used in repeated multiplication. The number to be multiplied is the base of the power, and the small raised number that tells how many times to use it is called the *exponent*. *See also* exponent.

power of 10   A number formed by multiplying several 10s. The first power of 10 is 10, the second power of 10 is 100, and the third power of 10 is 1,000.

prime factorization   A multiplication that uses only prime numbers and produces the original number as its product.

prime number   A whole number whose only factors are itself and 1.

principal   The amount of money borrowed or invested.

probability   The chance of an event occurring, expressed as the number of ways it can occur, divided by the number of total possible outcomes.

probability distribution   A list of all the possible outcomes and the probability of each.

product   The result of multiplying.

proper fraction   A fraction whose value is less than 1.

proportion   Two equal ratios. The means of a proportion are the two middle numbers, and the extremes are the first and last numbers.

quadrants   The four sections into which the coordinate plane is divided by the axes.

quadratic equation   An equation of the form $ax^2 + bx + c = 0$, in which a, b, and c are real numbers and a is not 0.

quadratic form   An equation that has the structure of a quadratic equation, even though it involves powers or expressions that do not fit the definition of a quadratic equation. The form can be made clearer by making a substitution of a single variable for a larger expression.

quadratic formula $x = \frac{-b \pm \sqrt{b^2 - 4ac}}{2a}$; a formula that produces the solutions of a quadratic equation, $ax^2 + bx + c = 0$, when the values of a, b, and c are substituted into the formula and simplified.

quadratic trinomial   An algebraic expression with three terms. It contains a term with the variable squared, a term with the variable to the first power, and a constant term.

quotient   The result of division.

radical   The symbol for the square root, from the Latin word *radix,* meaning "root."

radicand   The number or expression under a radical.

range   The set of all outputs, all of which are values of y.

rate   A comparison of two quantities in different units—for example, miles per hour or dollars per day.

ratio   A comparison of two numbers by division.

rational equation   An equation that contains one or more rational expressions.

rational exponent   A fraction, $\frac{p}{r}$, in which the numerator, p, indicates a power to which the base should be raised, and the denominator, r, denotes a root to be taken.

rational expression   A quotient of two polynomials, provided that the polynomial in the denominator is not 0.

rational numbers   The set of all numbers that can be written as the quotient, or ratio, of two integers.

rationalizing the denominator   The process of transforming an algebraic fraction so that no radicals remain in the denominator.

real numbers   The name given to the set of all rational numbers and all irrational numbers.

reciprocal   Two numbers whose product is 1. Each number is the reciprocal of the other.

reduced row echelon form   A form of an augmented matrix that has ones along the diagonal of the row-by-row square portion of the matrix, zeros elsewhere in those columns, and a final column containing the solution.

regression line   The line of best fit for a set of two-variable data.

relation   A pairing of numbers from one set, called the *domain,* with numbers from another set, called the *range.*

relatively prime   Two numbers for which the only factor they have in common is 1.

remainder   The number left over at the end of a division problem. It's the difference between the dividend and the product of the divisor and quotient.

residual   The distance between the observed value of y and the value of y which the line predicts for each x-value, when a line is fitted to data.

root   The opposite operation of a power; a way of undoing a power.

slope   The ratio of the number of spaces up or down to the number of spaces horizontally you must count to get from one point on a line to another.

solving an equation   Determining the value(s) of the variable for which that sentence is true.

square numbers   Numbers created by raising a number to the second power.

standard form   In which the terms of a polynomial are written in order from the highest degree to the lowest.

substitution method   A method of solving a system of equations by isolating one variable in one equation, and replacing that variable in the other equation with the expression it equals. The equation is solved for the remaining variable, and that value is substituted back to find the second variable.

subtrahend   The number you take away in a subtraction problem.

sum   The result of addition.

synthetic division   A technique for dividing a polynomial by a first-degree binomial, or for evaluating a polynomial for a particular value of the variable. It is based on the nested form of a polynomial: $ax^2 + bx + cx + d = x(x(x(a) + b) + c) + d$.

synthetic substitution   Synthetic division used to evaluate a function.

system of equations   A collection of two (or more) equations in two (or more) variables. A unique solution requires as many equations as variables.

system of inequalities   A set of two (or more) inequalities in two (or more) variables that together define a region containing all the points that simultaneously solve all the inequalities.

term   An algebraic expression made up of numbers, variables, or both that are connected only by multiplication.

test point method   The method of solving polynomial inequalities by finding the zeros and evaluating the function for values between zeros.

transformation   Any change made to a parent function to translate or slide it, reflect or flip it, and stretch or compress it.

tree diagram   A visual device to show the many different ways a set of choices can be made. Each branch, followed to its tip, represents one complete set of choices.

turning points   Points on the graph of a polynomial function at which the graph changes from increasing to decreasing or from decreasing to increasing.

unlike terms   Terms with different variables, such as x and y.

variable   A letter or symbol that takes the place of a number.

vertex   In an absolute value graph, the turning point or the point of the V of a graph. In a parent graph, the vertex is at the origin.

vertex form   The equation of a parabola, $y = a(x - h)^2 + k$, achieved by completing the square on $y = ax^2 + bx + c$, that shows the coordinates of the vertex, or turning point, of the parabola (h, k).

vertical asymptote   A vertical line at the point of discontinuity. The portions of the graph above and below the vertical asymptote either increase or decrease without bound as they get close to the vertical asymptote.

whole numbers   The set of numbers formed by adding a 0 to the natural numbers.

x-coordinate   The first number in an ordered pair that indicates horizontal movement.

y-coordinate   The second number in an ordered pair that indicates vertical movement.

zero exponent   Any nonzero number to the 0 power, which is 1.

Resources

One of the first things you learn as a teacher is that sometimes the best way to help a student understand an idea is to let someone else explain it. Teachers always try to find a way to make things clear, but sometimes you just have to get out of the way and let someone else try. So if you've read what I've written and still want to hear it from someone else, that's fine. You can, of course, search through a bookstore or type the subject into your favorite search engine, but here are a few resources you might find helpful.

Idiot's Guides

If you want a little background, here are some other *Idiot's Guides* you might want to look at.

Wheater, Carolyn. *The Complete Idiot's Guide to Basic Math and Pre-Algebra.* Indianapolis, IN: Alpha Books, 2014.

Wheater, Carolyn. *Idiot's Guides: Algebra I.* Indianapolis, IN: Alpha Books, 2015.

Gardner, Jane P. *The Complete Idiot's Guide to Algebra Practice Problems.* Indianapolis, IN: Alpha Books, 2011.

Fotiyeva, Izolda. *The Complete Idiot's Guide to Algebra Word Problems.* Indianapolis, IN: Alpha Books, 2010.

Online

There are many, many online resources. Some are good, some are great, and some are not. When you're looking for math or calculator help, as at any time you search the internet, don't immediately believe everything you see. Here are a few resources I like.

Math Help

Paul's Online Math Notes (http://tutorial.math.lamar.edu/Classes/Alg/Alg.aspx): While the main site focuses primarily on calculus and other college courses, there is an algebra section that's quite well done.

Mathforum.org: This site offers you a chance to "Ask Dr. Math" for help with algebra II or any other math. You can send them your question, but you're likely to find it has already been answered if you search the library of questions and answers.

Coolmath.com: This colorful site—which is so colorful it might not seem like a serious math site at first glance—is full of tutorials and study tips that you might find helpful, and it does try to keep it fun.

Khanacademy.org: This site has a large collection of video lessons that cover topics from basic arithmetic, through pre-algebra, and on to algebra and beyond.

VirtualNerd.com: This site has tutorials available for many algebra II topics. The listing of topics has fundamentals at the bottom and more advanced topics on top, but you can search if you're not sure where to find your question.

Ixl.com: If you'd like to quiz yourself, this site has practice problems on all the topics you'll encounter in algebra II, and they'll explain if you make a mistake.

Sophia.org (sophia.org/sophia-for-students): This site provides access to tutorials from many different authors, as well as quiz questions.

Calculator Help

Education.ti.com: As your graphing calculator becomes more important, you may want to make use of calculator instruction manuals available online. At this site, you can download a PDF version of the guidebook for the TI-84 calculator.

www2.stetson.edu: If you find the whole guidebook for the TI-84 overwhelming, this site has short sets of instructions to help you accomplish common tasks on your calculator.

Checkpoint Answers

Chapter 1

1. -13: integer, rational, real

2. 14.1: rational, real

3. $-\frac{10}{3}$: rational, real

4. 12: natural, integer, rational, real

5. $\sqrt{17}$: irrational, real

6. $6 - 8 + 4(11 - 3) \div 16 = 6 - 8 + 4(8) \div 16 = 6 - 8 + 32 \div 16 = 6 - 8 + 2 = 0$

7. $9x - 15x \div 3 - 7x + 13x = 9x - 5x - 7x + 13x = 10x$

8. $-4(7y - 8) + 11y - (5 - 6y) = -28y + 32 + 11y - 5 + 6y = -11y + 27$

9. $\frac{42x+18}{6} - \frac{18(12-28x)}{2^3} = \frac{42x+18}{6} - \frac{216-504x}{8} = 7x + 3 - \left(27 - 63x\right)$
$$= 7x + 3 - 27 + 63x = 70x - 24$$

10. $\frac{9\left(40x-75x^2\right)}{15x} = \frac{360x-675x^2}{15x} = 24 - 45x$

11. $4x - 2(3x - 8) = 12 - 7(3 - 4x)$
$4x - 6x + 16 = 12 - 21 + 28x$
$-2x + 16 = -9 + 28x$
$25 = 30x$
$x = \frac{5}{6}$

12. $2 - (7x + 5) = 13 - 3x$
$2 - 7x - 5 = 13 - 3x$
$-7x - 3 = 13 - 3x$
$-4x = 16$
$x = -4$

13. $3(4x + 2) = 8 - (2x + 12) + x - 3$
$12x + 6 = 8 - 2x - 12 + x - 3$
$12x + 6 = -x - 7$
$13x = -13$
$x = -1$

14. $2 - 5(x - 3) < 1 + x - 8$
$2 - 5x + 15 < x - 7$
$-5x + 17 < x - 7$
$-6x < -24$
$x > 4$

15. $x - (8 - 2x) \geq 6 + 5x - 18$
$x - 8 + 2x \geq 5x - 12$
$3x - 8 \geq 5x - 12$
$-2x \geq -4$
$x \leq 2$

16. $y = \frac{3}{5}x - 4$

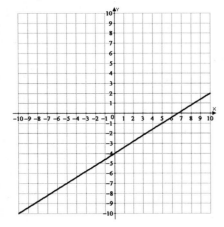

17. $y = -3x + 8$

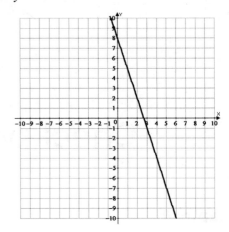

18. $6x + 15y = 30$

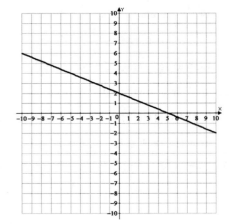

19. $9x - y = -9$

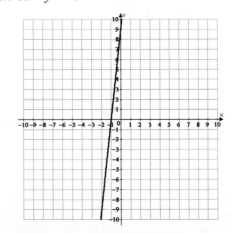

20. $5y - 20 = 4x$

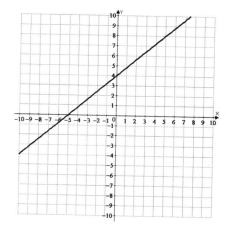

21. $y = \frac{1}{4}x - 5$

22. $y - (-1) = -2(x - 5) \Rightarrow y = -2x + 9$

23. $y - (-5) = \frac{2}{3}(x - (-6)) \Rightarrow y = \frac{2}{3}x - 1$

24. $m = \frac{5-1}{5-(-3)} = \frac{4}{8} = \frac{1}{2}$ and then
$y - 1 = \frac{1}{2}(x - (-3)) \Rightarrow y = \frac{1}{2}x + \frac{3}{2}$

25. $m = \frac{1-(-7)}{2-(-2)} = 2$ and then
$y - 1 = 2(x - 2) \Rightarrow y = 2x - 3$

26. $(x + 7)(x + 4) = x^2 + 11x + 28$

27. $(x - 3)(x - 5) = x^2 - 8x + 15$

28. $(x - 6)(x + 9) = x^2 + 3x - 54$

29. $(2x + 5)(3x - 7) = 6x^2 + x - 35$

30. $(4x - 7)(3x + 2) = 12x^2 - 13x - 14$

31. $x^2 - 6x + 5 = (x - 5)(x - 1)$

32. $x^2 + 11x + 30 = (x + 5)(x + 6)$

33. $x^2 + x - 20 = (x + 5)(x - 4)$

34. $6x^2 + 13x + 7 = (6x + 7)(x + 1)$

35. $15x^2 - 4x - 4 = (3x - 2)(5x + 2)$

Chapter 2

1. The relation is a function because each input (x) has only one output (y).

2. The relation $x^2 + 4y^2 = 16$ is not a function. When solved for y, it gives two y-values for each x: $x^2 + 4y^2 = 16 \rightarrow 4y^2 = 16 - x^2 \rightarrow y^2 = \frac{16-x^2}{4} \rightarrow y = \frac{\sqrt{16-x^2}}{2}$.

3. $f(-1) = 2(-1)^3 - 8(-1)^2 + (-1) - 5 = -2 - 8 - 1 - 5 = -16$

4. $g(2) = \sqrt{5(2)^2 - 4} = \sqrt{5 \cdot 4 - 4} = \sqrt{16} = 4$

5. $f(0) = \frac{2(0)-8}{0+1} = -8$

6. $f + g(x) = 3x - 7 + 9x^2 - 49 = 9x^2 + 3x - 56$, Domain: all real numbers

7. $g - f(x) = 9x^2 - 49 - (3x - 7) = 9x^2 - 3x - 42$, Domain: all real numbers

8. $f \cdot g(x) = (3x - 7)(9x^2 - 49) = 27x^3 - 63x^2 - 147x + 343$, Domain: all real numbers

9. $\frac{f}{g}(x) = \frac{3x-7}{9x^2-49} = \frac{3x-7}{(3x+7)(3x-7)} = \frac{1}{3x+7}$, Domain: $x \neq \pm\frac{7}{3}$

10. $\frac{g}{f}(x) = \frac{9x^2-49}{3x-7} = \frac{(3x+7)(3x-7)}{3x-7} = 3x + 7$, Domain: $x \neq \frac{7}{3}$

11. $f \circ g(x) = f(x^2 + 2) = \sqrt{x^2 + 2 - 2} = \sqrt{x^2} = |x|$, Domain: all real numbers

12. $g \circ f(x) = g(\sqrt{x-2}) = \sqrt{x-2}^2 + 2 = x - 2 + 2 = x$, Domain: $x \geq 2$

13. $f \circ h(x) = f(3x - 1) = \sqrt{3x - 1 - 2} = \sqrt{3x - 3}$, Domain: $x \geq 1$

14. $g \circ h(x) = g(3x - 1) = (3x - 1)^2 + 2 = 9x^2 - 6x + 1 + 2 = 9x^2 - 6x + 3$, Domain: all real numbers

15. $h(f(g(x))) = h(f(x^2 + 2)) = h(|x|) = 3|x| - 1$, Domain: all real numbers

16. $f(x) = 4x - 1$, $x = 4y - 1 \rightarrow x + 1 = 4y$, $f^{-1}(x) = \frac{x+1}{4}$. Domain: all real numbers

17. $g(x) = \sqrt{x^2 + 9}$, $x = \sqrt{y^2 + 9} \rightarrow x^2 = y^2 + 9 \rightarrow x^2 - 9 = y^2 \rightarrow \pm\sqrt{x^2 - 9} = y$. Because the domain of g is $x \leq 0$, $g^{-1}(x) = -\sqrt{x^2 - 9}$. Domain: $x \geq 3$.

18. $f(x) = 5 - x^2$, $x = 5 - y^2 \rightarrow y^2 = 5 - x \rightarrow y = \pm\sqrt{5 - x}$. Because the domain of f is $x \geq 0$, $f^{-1}(x) = \sqrt{5 - x}$. Domain: $x \leq 5$.

19. $g(x) = \frac{1}{x}$, $x = \frac{1}{y} \rightarrow xy = 1$, $g^{-1}(x) = \frac{1}{x}$. Domain: $x \neq 0$.

20. $f(x) = \frac{x+1}{x-1}$, $x = \frac{y+1}{y-1} \rightarrow xy - x = y + 1 \rightarrow xy - y = x + 1 \rightarrow y(x-1) = x + 1$, $f^{-1}(x) = \frac{x+1}{x-1}$. Domain: $x \neq 1$.

21. $g(x) = (x - 3)^2 + 2$

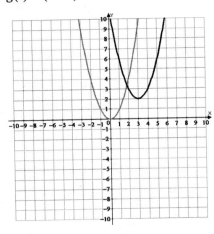

23. $g(x) = -\frac{1}{2}x^2 + 5$

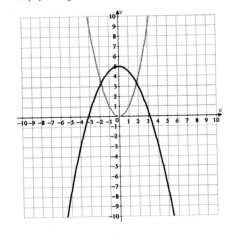

22. $g(x) = (x + 4)^2 - 1$

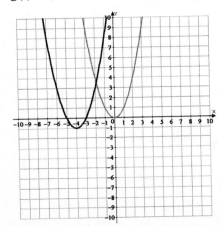

24. $g(x) = -(x + 5)^2 + 3$

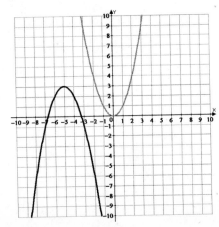

25. $g(x) = -2(x + 4)^2 + 7$

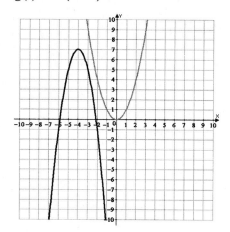

Chapter 3

1. $y \le \frac{2}{3}x - 5$

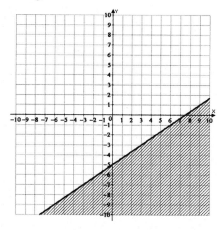

2. $y > -4x + 9$

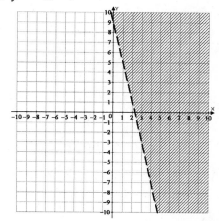

3. $3x - 2y < 12$

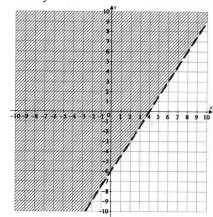

4. $2x + y \ge 8$

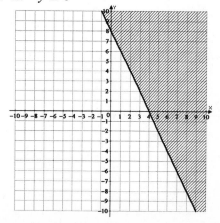

5. $3y < 7 - 2x$

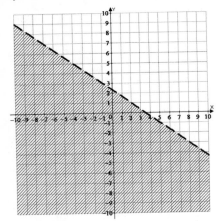

8. $y \approx 0.86x + 22.27$

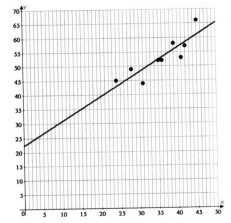

6. $y = -1.35x + 24.8$

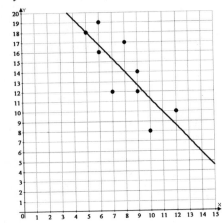

9. $y = 0.195x - 0.36$

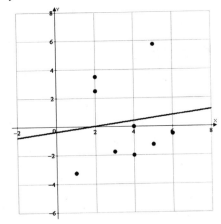

7. $y \approx 7.21x + 7.74$

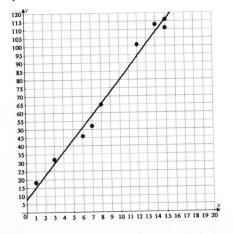

10. $y \approx -12.24x + 494.03$

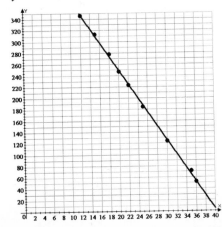

11. $f(x) = \begin{cases} 3x - 5 & \text{if } x \geq 2 \\ 9 - 2x & \text{if } x < 2 \end{cases}$

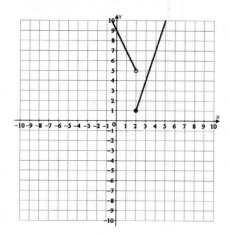

13. $y = \begin{cases} x + 4 & \text{if } x \geq 3 \\ x - 7 & \text{if } x < 3 \end{cases}$

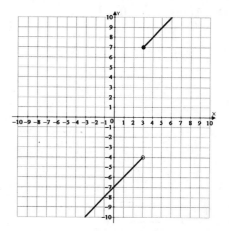

12. $g(x) = \begin{cases} \frac{1}{2}x + 5 & \text{if } x < 0 \\ -\frac{3}{5}x - 7 & \text{if } x \geq 0 \end{cases}$

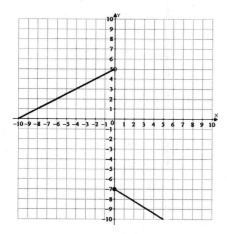

14. $f(x) = \begin{cases} 2x + 9 & \text{if } x < -2 \\ 5 & \text{if } -2 \leq x \leq 4 \\ x - 4 & \text{if } x > 4 \end{cases}$

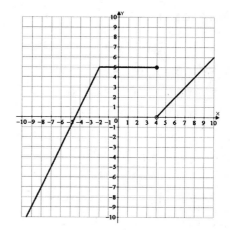

15. $g(x) = \begin{cases} -4 & \text{if } x \le -3 \\ -2 & \text{if } -3 < x \le -1 \\ 1 & \text{if } -1 < x \le 1 \\ 3 & \text{if } 1 < x \le 3 \\ 5 & \text{if } x > 3 \end{cases}$

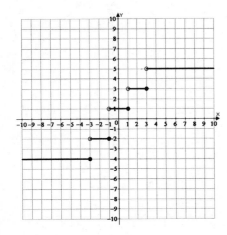

Chapter 4

1.
$$|4x - 3| = 9$$
$$4x - 3 = 9 \quad \text{or} \quad 4x - 3 = -9$$
$$4x = 12 \quad \text{or} \quad 4x = -6$$
$$x = 3 \quad \text{or} \quad x = -\tfrac{3}{2}$$

2.
$$|5 - 3x| + 7 = 25$$
$$|5 - 3x| = 18$$
$$5 - 3x = 18 \quad \text{or} \quad 5 - 3x = -18$$
$$-3x = 13 \quad \text{or} \quad -3x = -23$$
$$x = -\tfrac{13}{3} \quad \text{or} \quad x = \tfrac{23}{3}$$

3.
$$|8 - 3x| - 5 = 12$$
$$|8 - 3x| = 17$$
$$8 - 3x = 17 \quad \text{or} \quad 8 - 3x = -17$$
$$-3x = 9 \quad \text{or} \quad -3x = -25$$
$$x = -3 \quad \text{or} \quad x = \tfrac{25}{3}$$

4.
$$4 - 2|x + 1| = -18$$
$$-2|x + 1| = -22$$
$$|x + 1| = 11$$
$$x + 1 = 11 \quad \text{or} \quad x + 1 = -11$$
$$x = 10 \quad \text{or} \quad x = -12$$

5.
$$7|8x - 3| + 12 = 89$$
$$7|8x - 3| = 77$$
$$|8x - 3| = 11$$
$$8x - 3 = 11 \quad \text{or} \quad 8x - 3 = -11$$
$$8x = 14 \quad \text{or} \quad 8x = -8$$
$$x = \tfrac{7}{4} \quad \text{or} \quad x = -1$$

6.
$$|x + 7| > 3$$
$$x + 7 > 3 \quad \text{or} \quad x + 7 < -3$$
$$x > -4 \quad \text{or} \quad x < -10$$

7.
$$|2x + 5| - 7 > 22$$
$$|2x + 5| > 29$$
$$2x + 5 > 29 \quad \text{or} \quad 2x + 5 < -29$$
$$2x > 24 \quad \text{or} \quad 2x < -34$$
$$x > 12 \quad \text{or} \quad x < -17$$

8.
$$3 - |3 - 5x| \le 1$$
$$-|3 - 5x| \le -2$$
$$|3 - 5x| \ge 2$$
$$3 - 5x \ge 2 \quad \text{or} \quad 3 - 5x \le -2$$
$$-5x \ge -1 \quad \text{or} \quad -5x \le -5$$
$$x \le \tfrac{1}{5} \quad \text{or} \quad x \ge 1$$

9.
$$4 + |2x + 3| \geq 11$$
$$|2x + 3| \geq 7$$
$$2x + 3 \geq 7 \quad \text{or} \quad 2x + 3 \leq -7$$
$$2x \geq 4 \quad \text{or} \quad 2x \leq -10$$
$$x \geq 2 \quad \text{or} \quad x \leq -5$$

10.
$$|9 - 2x| + 7 < 20$$
$$|9 - 2x| < 13$$
$$9 - 2x < 13 \quad \text{and} \quad 9 - 2x > -13$$
$$-2x < 4 \quad \text{and} \quad -2x > -22$$
$$x > -2 \quad \text{and} \quad x < 11$$

11. $f(x) = 2|x - 4|$

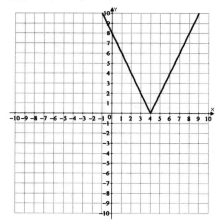

12. $g(x) = |x + 5| - 3$

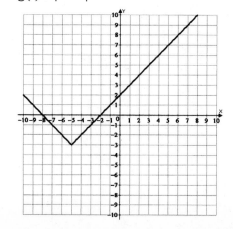

13. $y = 3 - |x - 2|$

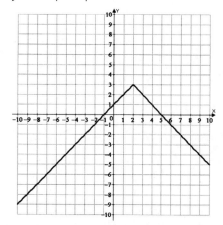

14. $f(x) = 5 - |x + 1|$

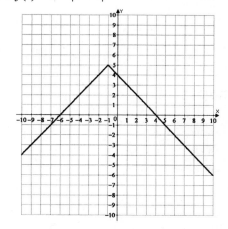

15. $g(x) = 3|x - 7| + 5$

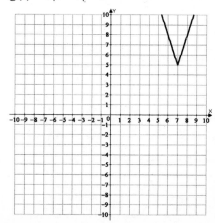

Chapter 5

1. $x = -2, y = 4$

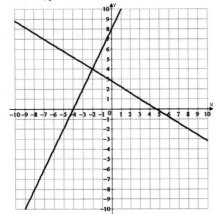

2. $x = 4, y = 7$

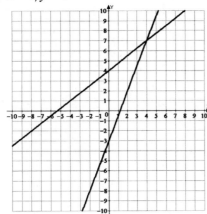

3. $x = -6, y = -7$

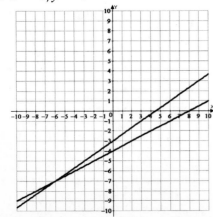

4. $x = -5, y = -8$

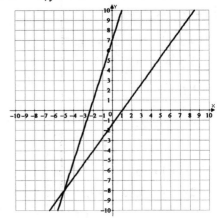

5. $x = 0, y = -2$

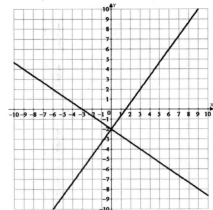

6. $y = 2x - 7$

$3x + (2x - 7) = 18$

$5x - 7 = 18$

$5x = 25$

$x = 5$

$y = 2(5) - 7 = 3$

7. $x + y = 8 \rightarrow y = 8 - x$

$3x - 7(8 - x) = -6$

$3x - 56 + 7x = -6$

$10x - 56 = -6$

$10x = 50$

$x = 5$

$5 + y = 8 \rightarrow y = 3$

8. $x - 3y = -7 \rightarrow x = 3y - 7$ $x - 3(-1) = -7$

$2y - 3x = 14$ $x + 3 = -7$

$2y - 3(3y - 7) = 14$ $x = -10$

$2y - 9y + 21 = 14$

$-7y = -7$

$y = -1$

9. $6x - y = 17 \rightarrow y = 6x - 17$

$5x + 2(6x - 17) = 34$ $6(4) - y = 17$

$5x + 12x - 34 = 34$ $24 - y = 17$

$17x = 68$ $y = 7$

$x = 4$

10. $4x - 2y = 42 \rightarrow 4x - 42 = 2y \rightarrow y = 2x - 21$

$3x + 5(2x - 21) = 12$ $4(9) - 2y = 42$

$3x + 10x - 105 = 12$ $36 - 2y = 42$

$13x = 117$ $-2y = 6$

$x = 9$ $y = -3$

11. $7x - 3y = 1$

$\underline{-3(2x - y) = (1)(-3)}$ $7(-2) + 3y = 1$

$7x - 3y = 1$ $-14 + 3y = 1$

$\underline{-6x + 3y = -3}$ $3y = 15$

$x = -2$ $y = 5$

12. $2x + 3y = 21$

$\underline{-3(4x + y) = (9)(-3)}$ $2(0.6) + 3y = 21$

$2x + 3y = 21$ $1.2 + 3y = 21$

$\underline{-12x - 3y = -27}$ $3y = 19.8$

$-10x = -6$ $y = 6.6$

$x = 0.6$

13. $3x - 7y = 30$

$\underline{7(5x + y) = (12)(7)}$

$3x - 7y = 30$ $3(3) - 7y = 30$

$\underline{35x + 7y = 84}$ $-7y = 21$

$38x = 114$ $y = -3$

$x = 3$

14. $-2(2x+4y)=(5)(-2)$

$$\frac{4x+5y=7}{-4x-8y=-10}$$

$$\frac{4x+5y=7}{-3y=-3}$$

$$y=1$$

$4x+5(1)=7$

$4x=2$

$x=\frac{2}{4}=\frac{1}{2}$

15. $8(2x+3y)=(4)(8)$

$$\frac{3(3x-8y)=(-9)(3)}{16x+24y=32}$$

$$\frac{9x-24y=-27}{25x=5}$$

$$x=0.2$$

$2(0.2)+3y=4$

$0.4+3y=4$

$3y=3.6$

$y=1.2$

16.

$$\frac{2x-y-3z=10}{x-2y+3z=-22}$$

$$3x-3y=-12$$

$x-2y+3z=-22$

$$\frac{3(3x+5y-z)=(63)(3)}{x-2y+3z=-22}$$

$$\frac{9x+15y-3z=189}{10x+13y=167}$$

$10(3x-3y)=(-12)(10)$

$$\frac{-3(10x+13y)=(167)(-3)}{30x-30y=-120}$$

$$\frac{-30x-39y=-501}{-69y=-621}$$

$$y=9$$

$3x-3(9)=-12$ $2(5)-9-3z=10$

$3x=15$ $-3z=9$

$x=5$ $z=-3$

17. $4x - 2y + 3z = -20$ $2x + 2y - z = 28$

$\underline{\quad 2x + 2y - z = 28 \quad}$ $\underline{\quad x - 2y + 3z = -32 \quad}$

$6x + 2z = 8$ $3x + 2z = -4$

$6x + 2z = 8$ $4x - 2y + 3z = -20$

$\underline{-2(3x + 2z) = (-4)(-2)}$ $3x + 2(-8) = -4$ $4(4) - 2y + 3(-8) = -20$

$6x + 2z = 8$ $3x - 16 = -4$ $16 - 2y - 24 = -20$

$\underline{-6x - 4z = 8}$ $3x = 12$ $-2y - 8 = -20$

$-2z = 16$ $x = 4$ $-2y = -12$

$z = -8$ $y = 6$

18. $2(3x - y + 4z) = (21)(2)$ $3(5x + 2y - 3z) = (-6)(3)$

$\underline{\quad 5x + 2y - 3z = -6 \quad}$ $\underline{\quad 5x - 6y + z = -2 \quad}$

$6x - 2y + 8z = 42$ $15x + 6y - 9z = -18$

$\underline{5x + 2y - 3z = -6}$ $\underline{5x - 6y + z = -2}$

$11x + 5z = 36$ $20x - 8z = -20$

$8(11x + 5z) = (36)(8)$

$\underline{5(20x - 8z) = (-20)(5)}$ $11(1) + 5z = 36$ $3(1) - y + 4(5) = 21$

$88x + 40z = 288$ $5z = 25$ $-y + 23 = 21$

$\underline{100x - 40z = -100}$ $z = 5$ $-y = -2$

$188x = 188$ $y = 2$

$x = 1$

19.
$$x - 7y - 3z = 6$$
$$-3(3x + y - z) = 8(-3)$$
$$x - 7y - 3z = 6$$
$$-9x - 3y + 3z = -24$$
$$-8x - 10y = -18$$

$$5(3x + y - z) = 8(5)$$
$$x - y + 5z = 12$$
$$15x + 5y - 5z = 40$$
$$x - y + 5z = 12$$
$$16x + 4y = 52$$

$$2(-8x - 10y) = -18(2)$$
$$16x + 4y = 52$$
$$-16x - 20y = -36$$
$$16x + 4y = 52$$
$$-16y = 16$$
$$y = -1$$
$$16x - 4 = 52$$
$$16x = 56$$
$$x = \tfrac{7}{2}$$
$$z = \tfrac{3}{2}$$

20.
$$x + y - z = -6$$
$$x - y + z = 14$$
$$2x = 8$$
$$x = 4$$

$$x - y + z = 14$$
$$x - y - z = 8$$
$$2x - 2y = 22$$
$$x - y = 11$$

$$4 - y = 11$$
$$y = -7$$
$$4 - 7 - z = -6$$
$$-z = -3$$
$$z = 3$$

21.
$$y - 2x \le 8$$
$$3x + 5y > 14$$

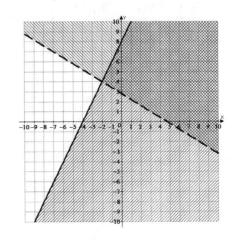

22. $2y < 5x - 6$

    $3x + 16 \le 4y$

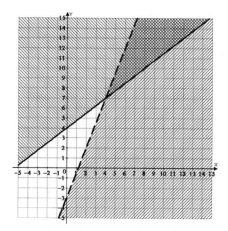

24. $y - 6 \le 3x + 1$

    $4x - 3y \ge 4$

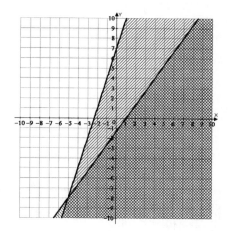

23. $2x - 3y > 9$

    $x - 2y < 8$

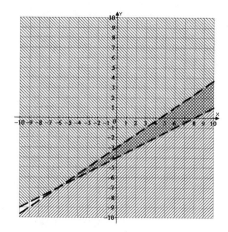

25. $y > -\frac{2}{3}x - 2$

    $4x - 3y \le 6$

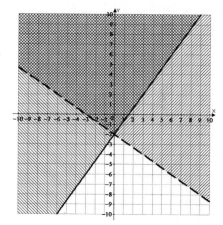

26. Maximum profit of $91 when $x = 2$ and $y = 3$.

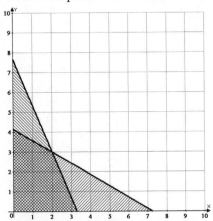

x	y	Profit = $14x + 21y$
0	0	0
$4\frac{1}{7}$	0	$14(4\frac{1}{7}) = 58$
0	$3\frac{2}{7}$	$21(3\frac{2}{7}) = 69$
2	3	$14(2) + 21(3) = 91$

27. Maximum profit of approximately $31.13 when $x = 3$ and $y = 2.5$.

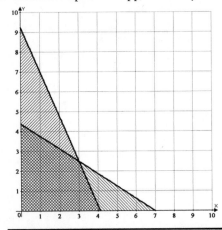

x	y	Profit = $7.25x + 3.75y$
0	0	0
0	4.75	$7.25(0) + 3.75(4.75) \approx 17.81$
$4\frac{1}{9}$	0	$7.25(4\frac{1}{9}) + 3.75(0) \approx 29.81$
3	2.5	$7.25(3) + 3.75(2.5) = 21.75 + 9.375 \approx 31.13$

28. Minimum cost of \$140 when $x = 2$ and $y = 3\frac{1}{3}$.

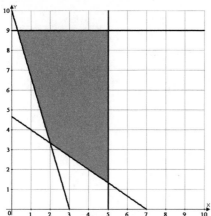

x	y	Cost = 20x + 30y
0.3	9	20(0.3) + 30(9) = 6 + 270 = 276
5	9	20(5) + 30(9) = 100 + 270 = 370
5	$1\frac{2}{3}$	20(5) + 30($1\frac{2}{3}$) = 100 + 50 = 150
2	$3\frac{1}{3}$	20(2) + 30($3\frac{1}{3}$) = 40 + 100 = 140

29. Minimum cost of \$360 when $x = 0$ and $y = 3$.

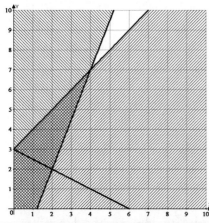

x	y	Cost = 150x + 120y
0	3	150(0) + 120(3)= 360
4	7	150(4) + 120(7) = 600 + 840 = 1,440
2	2	150(2) + 120(2)= 300 + 240 = 540

30. Maximum profit of $129 when $x = 7$ and $y = 3$.

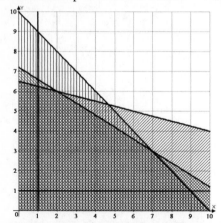

x	y	Profit = 12x + 15y
1	1	12 + 15 = 27
1	6.25	12(1) + 15(6.25) = 12 + 93.75 = 105.75
2	6	12(2) + 15(6) = 24 + 90 = 114
7	3	12(7) + 15(3)= 84 + 45 = 129
9	1	12(9) + 15(1) = 108 + 15 = 123

Chapter 6

1. $[A] + [F] = \begin{bmatrix} 3 & -3 \\ 0 & 3 \end{bmatrix}$

2. $[B] - [D]$ is not possible, because one matrix is 3-by-2, and the other is 2-by-3.

3. $[E] \cdot [B] = \begin{bmatrix} 10 & -10 \end{bmatrix}$

4. $[F] \cdot [D] = \begin{bmatrix} 9 & -4 & -10 \\ -1 & 9 & 5 \end{bmatrix}$

5. $[B] \cdot [C] = \begin{bmatrix} 3 & -2 \\ -1 & 2 \\ 1 & 0 \end{bmatrix} \cdot \begin{bmatrix} 3 \\ -5 \end{bmatrix} = \begin{bmatrix} 19 & -13 & 3 \end{bmatrix}$

6. $\begin{vmatrix} 4 & 2 \\ 9 & 6 \end{vmatrix} = 24 - 18 = 6$

7. $\begin{vmatrix} -5 & 6 \\ 5 & 4 \end{vmatrix} = -20 - 30 = -50$

8. $\begin{vmatrix} 6 & -9 \\ 7 & -10 \end{vmatrix} = -60 + 63 = 3$

9. $\begin{vmatrix} -1 & 2 & 1 \\ 2 & 0 & 0 \\ 3 & -4 & 2 \end{vmatrix} = (0 + 0 - 8) - (0 + 0 + 8) = -16$

10. $\begin{vmatrix} 8 & 0 & 3 \\ 6 & 7 & 2 \\ -3 & 8 & -4 \end{vmatrix} =$

   $(-224 + 0 + 144) - (-63 + 128 + 0) =$
   $-80 - 65 = -145$

11. $\begin{vmatrix} 2 & 1 \\ 4 & -3 \end{vmatrix} = -10,\ \begin{vmatrix} 4 & 1 \\ 13 & -3 \end{vmatrix} = -25,\ \begin{vmatrix} 2 & 4 \\ 4 & 13 \end{vmatrix} = 10,$

   $x = \frac{-25}{-10} = \frac{5}{2},\ y = \frac{10}{-10} = -1$

12. $\begin{vmatrix} 2 & 3 \\ 1 & 2 \end{vmatrix} = 1,$ $\begin{vmatrix} -8 & 3 \\ -3 & 2 \end{vmatrix} = -7,$ $\begin{vmatrix} 2 & -8 \\ 1 & -3 \end{vmatrix} = 2,$

$x = \frac{-7}{1},\ y = \frac{2}{1}$

13. $\begin{vmatrix} 3 & 7 \\ -1 & 1 \end{vmatrix} = 10,$ $\begin{vmatrix} -10 & 7 \\ 2 & 1 \end{vmatrix} = -24,$ $\begin{vmatrix} 3 & -10 \\ -1 & 2 \end{vmatrix} = -4,$

$x = \frac{-24}{10} = -2.4,\ y = \frac{-4}{10} = -0.4$

14. $\begin{vmatrix} 5 & 10 & 0 \\ 5 & 0 & 25 \\ 0 & 10 & 25 \end{vmatrix} = -2,500,$ $\begin{vmatrix} 70 & 10 & 0 \\ 270 & 0 & 25 \\ 300 & 10 & 25 \end{vmatrix} = -10,000,$

$\begin{vmatrix} 5 & 70 & 0 \\ 5 & 270 & 25 \\ 0 & 300 & 25 \end{vmatrix} = -12,500,$ $\begin{vmatrix} 5 & 10 & 70 \\ 5 & 0 & 270 \\ 0 & 10 & 300 \end{vmatrix} = -25,000,$

$x = \frac{-10,000}{-2,500} = 4,\ y = \frac{-12,500}{-2,500} = 5,\ z = \frac{-25,000}{-2,500} = 10$

15. $\begin{vmatrix} 2 & -3 & 1 \\ 1 & 3 & -2 \\ 4 & 5 & -3 \end{vmatrix} = 10,$ $\begin{vmatrix} 16 & -3 & 1 \\ -7 & 3 & -2 \\ -10 & 5 & -3 \end{vmatrix} = 14,$

$\begin{vmatrix} 2 & 16 & 1 \\ 1 & -7 & -2 \\ 4 & -10 & -3 \end{vmatrix} = -60,$ $\begin{vmatrix} 2 & -3 & 16 \\ 1 & 3 & -7 \\ 4 & 5 & -10 \end{vmatrix} = -48,$

$x = \frac{14}{10} = 1.4,\ y = \frac{-60}{10} = -6,\ z = \frac{-48}{10} = -4.8$

16. $\begin{bmatrix} 2 & -1 & 7 \\ 1 & 3 & -14 \end{bmatrix} \rightarrow \begin{bmatrix} 1 & 0 & 1 \\ 0 & 1 & -5 \end{bmatrix},\ x = 1, y = -5$

17. $\begin{bmatrix} 3 & 4 & -2 \\ 2 & -2 & 22 \end{bmatrix} \rightarrow \begin{bmatrix} 1 & 0 & 6 \\ 0 & 1 & -5 \end{bmatrix},\ x = 6, y = -5$

18. $\begin{bmatrix} 3 & -2 & 4 \\ 2 & 1 & 12 \end{bmatrix} \rightarrow \begin{bmatrix} 1 & 0 & 4 \\ 0 & 1 & 4 \end{bmatrix},\ x = 4, y = 4$

19. $\begin{bmatrix} 3 & 2 & 5 & -21 \\ 2 & 5 & 4 & 3 \\ -3 & -4 & 2 & -3 \end{bmatrix} \rightarrow \begin{bmatrix} 1 & 0 & 0 & -7 \\ 0 & 1 & 0 & 5 \\ 0 & 0 & 1 & -2 \end{bmatrix},\ x = -7, y = 5, z = -2$

20. $\begin{bmatrix} -3 & 4 & 0 & 38 \\ 5 & 0 & -3 & -42 \\ 0 & 3 & 2 & 23 \end{bmatrix} \rightarrow \begin{bmatrix} 1 & 0 & 0 & -6 \\ 0 & 1 & 0 & 5 \\ 0 & 0 & 1 & 4 \end{bmatrix},$

$x = -6, y = 5, z = 4$

21. $\begin{bmatrix} 3 & -1 \\ -5 & 2 \end{bmatrix}^{-1} = \begin{bmatrix} 2 & 1 \\ 5 & 3 \end{bmatrix}$

22. The inverse does not exist because the matrix is not square.

23. $\begin{bmatrix} 3 & -2 \\ 4 & 1 \end{bmatrix}^{-1} = \frac{1}{11}\begin{bmatrix} 1 & 2 \\ -4 & 3 \end{bmatrix} = \begin{bmatrix} \frac{1}{11} & \frac{2}{11} \\ -\frac{4}{11} & \frac{1}{11} \end{bmatrix}$

24. $\begin{bmatrix} 1 & 2 & 3 \\ -4 & 5 & 0 \\ 2 & 9 & 3 \end{bmatrix} = \begin{bmatrix} \frac{-15}{99} & \frac{-21}{99} & \frac{15}{99} \\ \frac{-12}{99} & \frac{3}{99} & \frac{12}{99} \\ \frac{46}{99} & \frac{5}{99} & \frac{-13}{99} \end{bmatrix} = \begin{bmatrix} \frac{-5}{33} & \frac{-7}{33} & \frac{5}{33} \\ \frac{-4}{33} & \frac{1}{33} & \frac{4}{33} \\ \frac{46}{99} & \frac{5}{99} & \frac{-13}{99} \end{bmatrix}$

25. $\begin{bmatrix} -3 & 4 & -2 \\ 0 & -1 & 5 \\ -2 & 0 & 3 \end{bmatrix} = \begin{bmatrix} \frac{3}{27} & \frac{12}{27} & \frac{-18}{27} \\ \frac{10}{27} & \frac{13}{27} & \frac{-15}{27} \\ \frac{2}{27} & \frac{8}{27} & \frac{-3}{27} \end{bmatrix} = \begin{bmatrix} \frac{1}{9} & \frac{4}{9} & \frac{-2}{3} \\ \frac{10}{27} & \frac{13}{27} & \frac{-5}{9} \\ \frac{2}{27} & \frac{8}{27} & \frac{-1}{9} \end{bmatrix}$

26. $\begin{bmatrix} 2 & -3 \\ 1 & -4 \end{bmatrix}\begin{bmatrix} x \\ y \end{bmatrix} = \begin{bmatrix} 11 \\ 3 \end{bmatrix},$ $\begin{bmatrix} x \\ y \end{bmatrix} = \begin{bmatrix} 2 & -3 \\ 1 & -4 \end{bmatrix}^{-1}\begin{bmatrix} 11 \\ 3 \end{bmatrix},$

$x = 7, y = 1$

27. $\begin{bmatrix} 3 & -4 \\ 1 & 2 \end{bmatrix}\begin{bmatrix} x \\ y \end{bmatrix} = \begin{bmatrix} -4 \\ 12 \end{bmatrix},$ $\begin{bmatrix} x \\ y \end{bmatrix} = \begin{bmatrix} 3 & -4 \\ 1 & 2 \end{bmatrix}^{-1}\begin{bmatrix} -4 \\ 12 \end{bmatrix},$

$x = 4, y = 4$

28. $\begin{bmatrix} 3 & -2 \\ 5 & 6 \end{bmatrix}\begin{bmatrix} x \\ y \end{bmatrix} = \begin{bmatrix} 5 \\ 13 \end{bmatrix},$ $\begin{bmatrix} x \\ y \end{bmatrix} = \begin{bmatrix} 3 & -2 \\ 5 & 6 \end{bmatrix}^{-1}\begin{bmatrix} 5 \\ 13 \end{bmatrix},$

$x = 2, y = \frac{1}{2}$

29.
$$\begin{bmatrix} 1 & 2 & 3 \\ 1 & -2 & 2 \\ 1 & 2 & -4 \end{bmatrix}\begin{bmatrix} x \\ y \\ z \end{bmatrix} = \begin{bmatrix} 14 \\ 3 \\ -7 \end{bmatrix}, \begin{bmatrix} x \\ y \\ z \end{bmatrix} = \begin{bmatrix} 1 & 2 & 3 \\ 1 & -2 & 2 \\ 1 & 2 & -4 \end{bmatrix}^{-1}\begin{bmatrix} 14 \\ 3 \\ -7 \end{bmatrix},$$
$x = 1, y = 2, z = 3$

30.
$$\begin{bmatrix} 1 & -1 & 1 \\ 1 & 2 & 3 \\ 1 & -4 & 5 \end{bmatrix}\begin{bmatrix} x \\ y \\ z \end{bmatrix} = \begin{bmatrix} 15 \\ 100 \\ 50 \end{bmatrix}, \begin{bmatrix} x \\ y \\ z \end{bmatrix} = \begin{bmatrix} 1 & -1 & 1 \\ 1 & 2 & 3 \\ 1 & -4 & 5 \end{bmatrix}^{-1}\begin{bmatrix} 15 \\ 100 \\ 50 \end{bmatrix}, x = 10, y = 15, z = 20$$

Chapter 7

1. $8^{\frac{5}{3}} = 32$

2. $(-32)^{\frac{2}{5}} = (-2)^2 = 4$

3. $\left(\frac{16}{25}\right)^{\frac{1}{2}} = \frac{4}{5}$

4. $10,000^{\frac{3}{4}} = 10^3 = 1,000$

5. $(-343)^{\frac{4}{3}} = 7^4 = 2,401$

6. $\sqrt[3]{8x^4y^5} = 2xy\sqrt[3]{xy^2}$

7. $\sqrt{27a^5b^4c^7} = 3a^2b^2c^3\sqrt{3ac}$

8. $\left(\frac{81x^7y^5}{16z^9}\right)^{\frac{1}{3}} = \left(\frac{3^3 \cdot 3 \cdot x^6 \cdot x \cdot y^3 \cdot y^2}{2^3 \cdot 2 \cdot z^9}\right)^{\frac{1}{3}} = \frac{3x^2y}{2z^3}\left(\frac{3}{2}xy^2\right)^{\frac{1}{3}}$

9. $\sqrt[5]{\frac{64x^7y^8}{z^{10}}} = \sqrt[5]{\frac{32 \cdot 2 \cdot x^5 \cdot x^2 y^5 \cdot y^3}{z^{10}}} = \frac{2xy\sqrt[5]{2x^2y^3}}{z^2}$

10.
$$\left(\frac{27x^3y^4}{8z^3}\right)^{\frac{5}{2}} = \left(\frac{3^3x^3y^4}{2^3z^3}\right)^{\frac{5}{2}} = \left(\frac{3^{\frac{15}{2}}x^{\frac{15}{2}}y^{10}}{2^{\frac{15}{2}}z^{\frac{15}{2}}}\right) = \left(\frac{3^7x^7y^{10}}{2^7z^7}\right)\left(\frac{3x}{2z}\right)^{\frac{1}{2}}$$
$$= \frac{2,187x^7y^{10}}{128z^7}\frac{\sqrt{3x}}{\sqrt{2z}} \cdot \frac{\sqrt{2z}}{\sqrt{2z}} = \frac{2,187x^7y^{10}}{128z^7}\frac{\sqrt{6xz}}{2z} = \frac{2,187x^7y^{10}\sqrt{6xz}}{256z^8}$$

11. $\frac{5}{2+\sqrt{3}} \cdot \frac{2-\sqrt{3}}{2-\sqrt{3}} = \frac{10-5\sqrt{3}}{4-3} = 10 - 5\sqrt{3}$

12. $\frac{3}{\sqrt{2}-9} \cdot \frac{\sqrt{2}+9}{\sqrt{2}+9} = \frac{3\sqrt{2}+27}{2-81} = \frac{3\sqrt{2}+27}{79}$

13. $\frac{2+3\sqrt{5}}{8-\sqrt{5}} \cdot \frac{8+\sqrt{5}}{8+\sqrt{5}} = \frac{16+2\sqrt{5}+24\sqrt{5}+15}{64-5} = \frac{31+26\sqrt{5}}{59}$

14. $\frac{2}{\sqrt[3]{5}} \cdot \left(\frac{\sqrt[3]{5}}{\sqrt[3]{5}}\right)^2 = \frac{2\sqrt[3]{25}}{(\sqrt[3]{5})^3} = \frac{2\sqrt[3]{25}}{5}$

15. $\frac{2+\sqrt[3]{6}}{\sqrt[3]{3}} \cdot \left(\frac{\sqrt[3]{3}}{\sqrt[3]{3}}\right)^2 = \frac{2\sqrt[3]{9}+\sqrt[3]{54}}{3} = \frac{2\sqrt[3]{9}+3\sqrt[3]{2}}{3}$

16. $\sqrt{8x-4} = 10 - 2\sqrt{x-1}$
$8x-4 = 100 - 40\sqrt{x-1} + 4(x-1)$
$4x-100 = -40\sqrt{x-1}$
$16x^2 - 800x + 10,000 = 1,600(x-1)$
$16x^2 - 2,400x + 11,600 = 0$
$x^2 - 150x + 725 = 0$
$x = 145$ (extraneous) or $x = 5$

17. $\sqrt[3]{4+5x} + 7 = 11$
$\sqrt[3]{4+5x} = 4$
$4+5x = 64$
$5x = 60$
$x = 12$

18. No solution.
$\sqrt{2x-5} = 1 - \sqrt{3x-5}$
$2x-5 = 1 - 2\sqrt{3x-5} + 3x - 5$
$2\sqrt{3x-5} = x + 1$
$4(3x-5) = x^2 + 2x + 1$
$x^2 - 10x + 21 = 0$
$x = 3$, $x = 7$ both extraneous

19. $\sqrt[3]{48-3x}+3=6$

$\sqrt[3]{48-3x}=3$

$48-3x=27$

$-3x=-21$

$x=7$

20. $\sqrt[4]{8x^2-15}=x$

$8x^2-15=x^4$

$x^4-8x^2+15=0$

$\left(x^2-5\right)\left(x^2-3\right)=0$

$x=\sqrt{5},\ x=\sqrt{3}$

$(x=-\sqrt{5},\ x=-\sqrt{3}$ are extraneous)

21. $f\left(x\right)=\sqrt{x-5}+3$

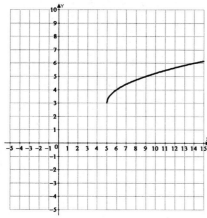

22. $y=2-3\sqrt{x+4}$

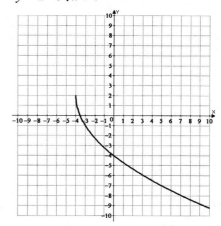

23. $f\left(x\right)=\frac{1}{2}\sqrt{4-x}-5$

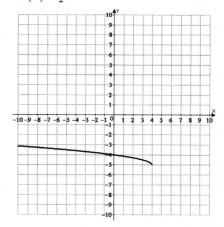

24. $g\left(x\right)=\sqrt[3]{x+4}-5$

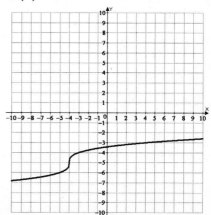

25. $y = 5 - 2\sqrt[3]{x+4}$

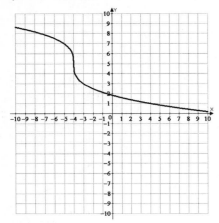

Chapter 8

1. $5x^2 = 80$

$x^2 = 16$

$x = \pm 4$

2. $\frac{1}{2}x^2 = \frac{25}{8}$

$x^2 = \frac{25}{4}$

$x = \pm\frac{5}{2}$

3. $(3x-7)^2 = 25$

$3x - 7 = \pm 5$

$3x - 7 = 5 \qquad 3x - 7 = -5$

$3x = 12 \qquad 3x = 2$

$x = 4 \qquad x = \frac{2}{3}$

4. $(4-5x)^2 = 48$

$4 - 5x = \pm 4\sqrt{3}$

$-5x = -4 \pm 4\sqrt{3}$

$x = \frac{4 \pm 4\sqrt{3}}{5}$

5. $(9+3y)^2 = 225$

$9 + 3y = \pm 15$

$9 + 3y = 15 \qquad 9 + 3y = -15$

$3y = 6 \qquad 3y = -24$

$y = 2 \qquad y = -8$

6. $x^2 - 15x + \left(-\frac{15}{2}\right)^2 = -2 + \left(-\frac{15}{2}\right)^2$

$\left(x - \frac{15}{2}\right)^2 = \frac{217}{4}$

$x - \frac{15}{2} = \pm\frac{\sqrt{217}}{2}$

$x = \frac{15 \pm \sqrt{217}}{2}$

7. $y^2 - 72y + (36)^2 = -8 + 1{,}296$

$(y - 36)^2 = 1{,}288$

$y - 36 = \pm\sqrt{1{,}288}$

$y = 36 \pm 2\sqrt{322}$

8. $2x^2 + 12x = 3$

$x^2 + 6x + 9 = \frac{3}{2} + 9$

$(x+3)^2 = \frac{21}{2}$

$x + 3 = \pm\sqrt{\frac{21}{2}} = \frac{\sqrt{42}}{2}$

$x = \frac{-6 \pm \sqrt{42}}{2}$

9. $3x^2 - 12x = -1$

$x^2 - 4x + 4 = \frac{-1}{3} + 4$

$(x-2)^2 = \frac{11}{3}$

$x - 2 = \pm\sqrt{\frac{11}{3}} = \pm\frac{\sqrt{33}}{3}$

$x = \frac{6 \pm \sqrt{33}}{3}$

10. $5x^2 + 13x = -2$

$$x^2 + \tfrac{13}{5}x + \left(\tfrac{13}{10}\right)^2 = -\tfrac{2}{5} + \left(\tfrac{13}{10}\right)^2$$

$$\left(x + \tfrac{13}{10}\right)^2 = \tfrac{129}{100}$$

$$x + \tfrac{13}{10} = \pm\tfrac{\sqrt{129}}{10}$$

$$x = \tfrac{-13\pm\sqrt{129}}{10}$$

11. $2x^2 + 3 = 12x$

$$x = \tfrac{-b\pm\sqrt{b^2-4ac}}{2a} = \tfrac{12\pm\sqrt{144-4(2)(3)}}{2(2)} = \tfrac{12\pm\sqrt{120}}{4} = \tfrac{12\pm2\sqrt{30}}{4} = \tfrac{6\pm\sqrt{30}}{2}$$

12. $2 = 4x - x^2$

$$x = \tfrac{-b\pm\sqrt{b^2-4ac}}{2a} = \tfrac{-4\pm\sqrt{16-4(-1)(-2)}}{2(-1)} = \tfrac{-4\pm\sqrt{8}}{-2} = \tfrac{-4\pm2\sqrt{2}}{-2} = 2\pm\sqrt{2}$$

13. $30z^2 - 17z = 0$

$$z = \tfrac{-b\pm\sqrt{b^2-4ac}}{2a} = \tfrac{17\pm\sqrt{289-4(30)(0)}}{2(30)} = \tfrac{17\pm17}{60}$$

$$z = 0, \quad z = \tfrac{34}{60} = \tfrac{17}{30}$$

14. $8x^2 - 9x - 2 = 0$

$$x = \tfrac{-b\pm\sqrt{b^2-4ac}}{2a} = \tfrac{9\pm\sqrt{81-4(8)(-2)}}{2(8)} = \tfrac{9\pm\sqrt{145}}{16}$$

15. $7x^2 - 7x + 1 = 0$

$$x = \tfrac{-b\pm\sqrt{b^2-4ac}}{2a} = \tfrac{7\pm\sqrt{49-4(7)(1)}}{2(7)} = \tfrac{7\pm\sqrt{21}}{14}$$

16. $b^2 - 4ac = 49 - 4(4)(1) = 49 - 16 = 33 > 0$. $4x^2 + 7x + 1 = 0$ has two real, irrational zeros.

17. $b^2 - 4ac = 9 - 4(2)(-3) = 9 + 24 = 33 > 0$. $2x^2 - 3x - 3 = 0$ has two real, irrational zeros.

18. $b^2 - 4ac = 4 - 4(5)(3) = 4 - 60 = -56 < 0$. $5x^2 + 2x + 3 = 0$ has two complex zeros.

19. $b^2 - 4ac = 144 - 4(9)(4) = 0$. $9x^2 + 12x + 4 = 0$ has one real zero (with a multiplicity of 2).

20. $b^2 - 4ac = 0 - 4(25)(-36) = 3{,}600 > 0$. $25x^2 - 36 = 0$ has two real, rational zeros.

21. $(3 - 2i) + (4 + 9i) - (2 - 5i) = (7 + 7i) - (2 - 5i) = 5 + 12i$

22. $(2 - 5i)(3 + 4i) = 6 + 8i - 15i + 20 = 26 - 7i$

23. $\tfrac{6-5i}{3+2i} \cdot \tfrac{3-2i}{3-2i} = \tfrac{18-12i-15i-10}{9+4} = \tfrac{8-27i}{13}$

24. $x^2 + 4x + 7 = 0$

$x = \frac{-b \pm \sqrt{b^2 - 4ac}}{2a} = \frac{-4 \pm \sqrt{16 - 4(1)(7)}}{2(1)} = \frac{-4 \pm \sqrt{-12}}{2} = \frac{-4 \pm 2i\sqrt{3}}{2} = -2 \pm i\sqrt{3}$

25. $3x^2 + 8x + 12 = 0$

$x = \frac{-b \pm \sqrt{b^2 - 4ac}}{2a} = \frac{-8 \pm \sqrt{64 - 4(3)(12)}}{2(3)} = \frac{-8 \pm \sqrt{-80}}{6} = \frac{-8 \pm 4i\sqrt{5}}{6} = \frac{-4 \pm 2i\sqrt{5}}{3}$

26. $f(x) = (x - 4)^2 + 3$

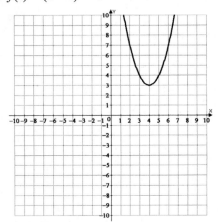

28. $g(x) = -3(x + 1)^2 - 5$

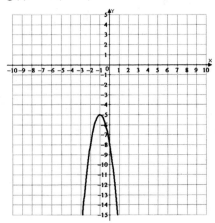

27. $y = 2(x + 5)^2 - 3$

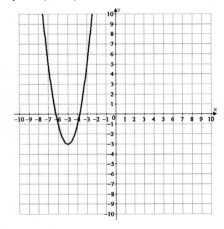

29. $y = -\frac{1}{2}(x - 4)^2 + 1$

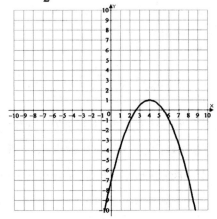

30. $f(x) = 4(x+3)^2 - 7$

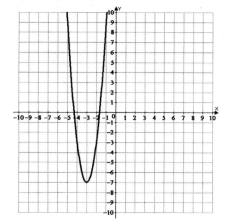

31. $y = a(x+3)^2 + 7$

$-41 = a(1+3)^2 + 7 \rightarrow a = -3$

$y = -3(x+3)^2 + 7$

32. $y = a(x-4)^2 - 1$

$1 = a(2-4)^2 - 1 \rightarrow a = \frac{1}{2}$

$y = \frac{1}{2}(x-4)^2 - 1$

33. $y = a(x+5)^2 - 4$

$23 = a(-2+5)^2 - 4 \rightarrow a = 3$

$y = 3(x+5)^2 - 4$

34. $17 = a(-2)^2 + b(-2) + c \rightarrow 4a - 2b + c = 17$

$-1 = a(1)^2 + b(1) + c \rightarrow a + b + c = -1$

$9 = a(6)^2 + b(6) + c \rightarrow 36a + 6b + c = 9$

$$\begin{bmatrix} 4 & -2 & 1 & 17 \\ 1 & 1 & 1 & -1 \\ 36 & 6 & 1 & 9 \end{bmatrix} \rightarrow \begin{bmatrix} 1 & 0 & 0 & 1 \\ 0 & 1 & 0 & -5 \\ 0 & 0 & 1 & 3 \end{bmatrix}$$

$y = 1x^2 - 5x + 3$

35. $-18 = a(-2)^2 + b(-2) + c \rightarrow 4a - 2b + c = -18$

$-5 = a(-1)^2 + b(-1) + c \rightarrow a - b + c = -5$

$10 = a(2)^2 + b(2) + c \rightarrow 4a + 2b + c = 10$

$$\begin{bmatrix} 4 & -2 & 1 & -18 \\ 1 & -1 & 1 & -5 \\ 4 & 2 & 1 & 10 \end{bmatrix} \rightarrow \begin{bmatrix} 1 & 0 & 0 & -2 \\ 0 & 1 & 0 & 7 \\ 0 & 0 & 1 & 4 \end{bmatrix}$$

$y = -2x^2 + 7x + 4$

Chapter 9

1. $y^2 + 8y + 16 = -4x - 24 + 16$

$(y+4)^2 = -4(x+2)$

$x + 2 = -\frac{1}{4}(y+4)^2$

2. $8y - 20 = x^2 + 4x$

$8y - 20 + \left(\frac{4}{2}\right)^2 = x^2 + 4x + \left(\frac{4}{2}\right)^2$

$8y - 20 + 4 = x^2 + 4x + 4$

$8(y-2) = (x+2)^2$

$y - 2 = \frac{1}{8}(x+2)^2$

3. $3y^2 - 6y = -9x + 24$

$\frac{3y^2 - 6y}{3} = \frac{-9x+24}{3}$

$y^2 - 2y = -3x + 8$

$y^2 - 2y + \left(\frac{-2}{2}\right)^2 = -3x + 8 + \left(\frac{-2}{2}\right)^2$

$(y-1)^2 = -3(x-3)$

$x - 3 = -\frac{1}{3}(y-1)^2$

4. $5y - 20 = 10x^2 + 40x$

   $y - 4 = 2x^2 + 8x$

   $\frac{y-4}{2} = x^2 + 4x$

   $\frac{y-4}{2} + \left(\frac{4}{2}\right)^2 = x^2 + 4x + \left(\frac{4}{2}\right)^2$

   $\frac{y-4}{2} + 4 = (x+2)^2$

   $\frac{y-4}{2} + \frac{8}{2} = (x+2)^2$

   $\frac{y+4}{2} = (x+2)^2$

   $y + 4 = 2(x+2)^2$

5. $7y^2 - 14y = -x + 2$

   $y^2 - 2y + 1 = \frac{-x+2}{7} + \frac{7}{7}$

   $(y-1)^2 = \frac{x-9}{-7}$

   $-7(y-1)^2 = x - 9$

6. $y = \frac{1}{2}(x-2)^2 + 5$

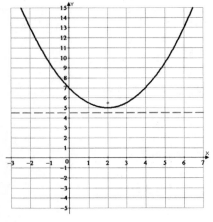

7. $x - 7 = -3(y+1)^2$

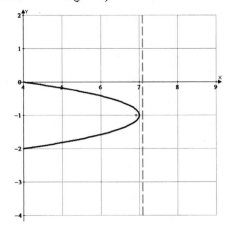

8. $y = -5(x+4)^2 - 7$

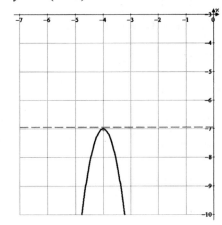

9. $x + 3 = -4(y-3)^2$

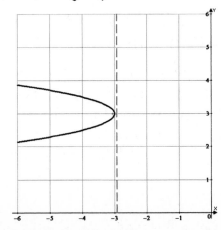

10. $x - 5 = -\frac{1}{4}(y + 3)^2$

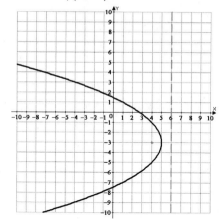

13. $(x + 4)^2 + (y - 1)^2 = 16$

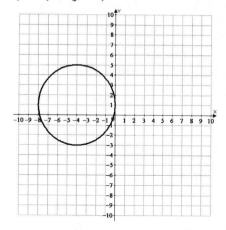

11. $(x + 5)^2 + (y - 2)^2 = 4$

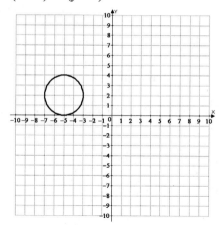

14. $x^2 + (y + 5)^2 = 49$

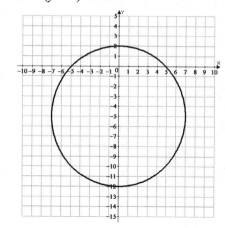

12. $(x - 3)^2 + (y - 1)^2 = 25$

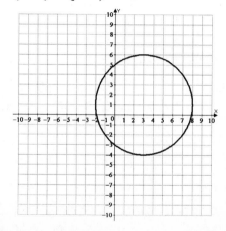

15. $(x - 2)^2 + y^2 = 9$

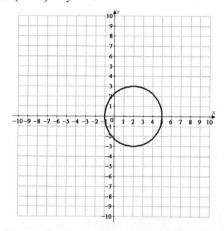

16. $x^2 + 4x + y^2 + 6y = 3$

$x^2 + 4x + 4 + y^2 + 6y + 9 = 3 + 4 + 9$

$(x + 2)^2 + (y + 3)^2 = 16$

17. $121x^2 + 25(y^2 - 6y) = 2{,}800$

$121x^2 + 25(y^2 - 6y + 9) = 2{,}800 + 9 \cdot 25$

$121x^2 + 25(y - 3)^2 = 3{,}025$

$\frac{121x^2}{3{,}025} + \frac{25(y-3)^2}{3{,}025} = 1$

$\frac{x^2}{25} + \frac{(y-3)^2}{121} = 1$

18. $3x^2 - 36x + 5y^2 + 30y = -138$

$3(x^2 - 12x + 36) + 5(y^2 + 6y + 9) = -138 + 3 \cdot 36 + 5 \cdot 9$

$3(x - 6)^2 + 5(y + 3)^2 = 15$

$\frac{(x-6)^2}{5} + \frac{(y+3)^2}{3} = 1$

19. $2x^2 + 16x + 3y^2 - 30y = -101$

$2(x^2 + 8x + 16) + 3(y^2 - 10y + 25) = -101 + 2 \cdot 16 + 3 \cdot 25$

$2(x + 4)^2 + 3(y - 5)^2 = 6$

$\frac{(x+4)^2}{3} + \frac{(y-5)^2}{2} = 1$

20. $4x^2 + 8x + 25y^2 - 50y = 71$

$4(x^2 + 2x + 1) + 25(y^2 - 2y + 1) = 71 + 4 \cdot 1 + 25 \cdot 1$

$4(x + 1)^2 + 25(y - 1)^2 = 100$

$\frac{(x+1)^2}{25} + \frac{(y-1)^2}{4} = 1$

21. $\frac{(x-3)^2}{16} + \frac{(y-2)^2}{25} = 1$

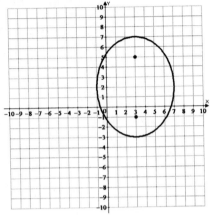

22. $\frac{(x-4)^2}{9} + \frac{y^2}{4} = 1$

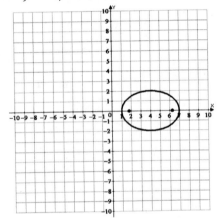

23. $\frac{(x-1)^2}{16} + \frac{(y-2)^2}{4} = 1$

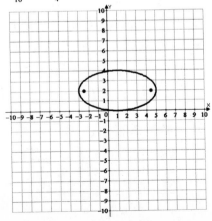

24. $(x+5)^2 + \frac{(y-1)^2}{4} = 1$

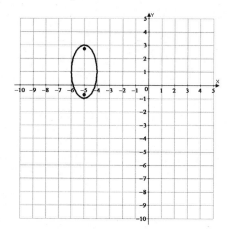

25. $c^2 = a^2 - b^2 \rightarrow 9 = 25 - b^2 \rightarrow b = 4$,

Equation: $\frac{x^2}{25} + \frac{y^2}{16} = 1$

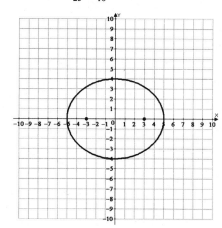

26. $x^2 - 2x + 1 - (y^2 + 4y + 4) = 28 + 1 - 4$

$(x-1)^2 - (y+2)^2 = 25$

$\frac{(x-1)^2}{25} - \frac{(y+2)^2}{25} = 1$

27. $9(x^2 + 2x) - 4(y^2 + 4y) = 43$

$9(x^2 + 2x + 1) - 4(y^2 + 4y + 4) = 43 + 9 - 16$

$9(x+1)^2 - 4(y+2)^2 = 36$

$\frac{(x+1)^2}{4} - \frac{(y+2)^2}{9} = 1$

28. $16y^2 - 64y - 9x^2 + 18x = 89$

$16(y^2 - 4y) - 9(x^2 - 2x) = 89$

$16(y^2 - 4y + 4) - 9(x^2 - 2x + 1) = 89 + 64 - 9$

$16(y-2)^2 - 9(x-1)^2 = 144$

$\frac{(y-2)^2}{9} - \frac{(x-1)^2}{16} = 1$

29. $x^2 + 4x - 25y^2 - 200y = 421$

$(x^2 + 4x + 4) - 25(y^2 - 8y + 16) = 421 + 4 - 400$

$(x+2)^2 - 25(y-4)^2 = 25$

$\frac{(x+2)^2}{25} - \frac{(y-4)^2}{1} = 1$

30. $25x^2 - (y^2 - 4y + 4) = 25$

$25x^2 - (y-2)^2 = 25$

$\frac{x^2}{1} - \frac{(y-2)^2}{25} = 1$

31. $\frac{y^2}{4} - \frac{x^2}{36} = 1$

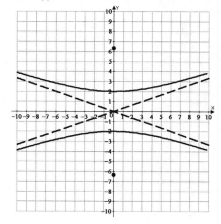

32. $\dfrac{(y-2)^2}{36} - \dfrac{(x+2)^2}{4} = 1$

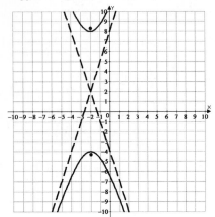

35. Equation: $\dfrac{y^2}{4} - \dfrac{x^2}{9} = 1$

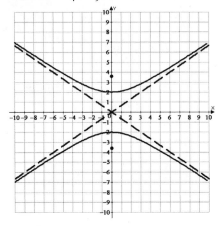

33. $\dfrac{(x+5)^2}{9} - \dfrac{(y-3)^2}{25} = 1$

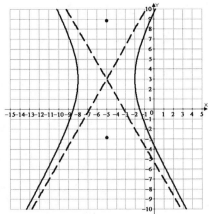

36.

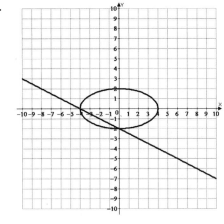

34. $(x-2)^2 - 4(y-1)^2 = 4$

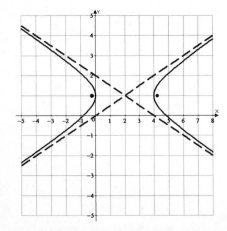

$$x^2 + 4y^2 = 16$$
$$x + 2y = -4 \rightarrow 2y = -x - 4$$
$$\rightarrow 4y^2 = (-x-4)^2 \rightarrow 4y^2 = x^2 + 8x + 16$$
$$x^2 + x^2 + 8x + 16 = 16$$
$$2x^2 + 8x = 2x(x+4) = 0$$

$$2x = 0 \qquad\qquad x + 4 = 0$$
$$x = 0 \qquad\qquad x = -4$$
$$x = 0, y = -2 \qquad x = -4, y = 0$$

37.

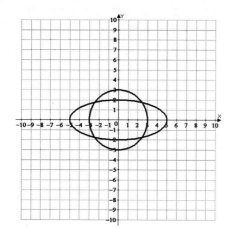

$$-25\left(x^2+y^2\right)=9(-25)$$
$$\underline{4x^2+25y^2=100}$$
$$\underline{-25x^2-25y^2=-225}$$
$$\underline{4x^2+25y^2=100}$$
$$-21x^2=-125$$
$$x^2=\tfrac{125}{21}$$
$$x=\tfrac{\pm5\sqrt{105}}{21}\approx\pm2.44$$

$$4\left(\tfrac{125}{21}\right)+25y^2=100$$
$$25y^2=100-\tfrac{500}{21}$$
$$y^2=\tfrac{64}{21}$$
$$y=\tfrac{\pm8\sqrt{21}}{21}\approx\pm1.75$$

38.

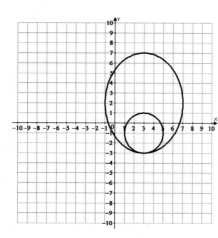

$$25(x-3)^2+16(y-2)^2=400$$
$$(x-3)^2+\left(y+1\right)^2=4\rightarrow(x-3)^2=4-\left(y+1\right)^2$$
$$25\left(4-\left(y+1\right)^2\right)+16(y-2)^2=400$$
$$100-25\left(y+1\right)^2+16(y-2)^2=400$$
$$-25\left(y^2+2y+1\right)+16\left(y^2-4y+4\right)=300$$
$$-9y^2-114y+39=300$$
$$-9y^2-114y-261=0$$
$$y=-9\tfrac{2}{3}\qquad\qquad y=-3$$
$$reject\qquad\qquad\quad x=3$$

39.

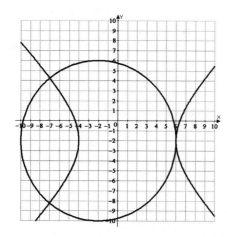

$$x^2-y^2-2x-4y=28$$
$$\underline{x^2+y^2+4x+4y=56}$$
$$2x^2+2x=84$$
$$x^2+x-42=0$$

$$x-6=0\qquad\qquad x+7=0$$
$$x=6\qquad\qquad\quad x=-7$$
$$36+y^2+24+4y=56\qquad 49+y^2-28+4y=56$$
$$y^2+4y+4=0\qquad\qquad y^2+4y-35=0$$
$$y=-2\qquad\qquad\qquad y\approx4.24, y\approx-8.24$$

40.

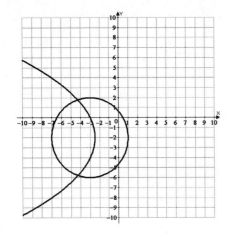

$$y^2 + 8x + 4y = -24$$
$$\underline{y^2 + 6x + 4y = 3 - x^2}$$
$$2x = -27 + x^2$$
$$x^2 - 2x - 27 = 0$$
$$x \approx 6.29 \quad x \approx -4.29$$
reject $\quad y \approx 1.79, y \approx -5.79$

Chapter 10

1. $t^3 + 30t^2 + 300t + 1{,}000 = (t + 10)^3$

2. $8x^3 - 125 = (2x - 5)(4x^2 + 10x + 25)$

3. $z^3 + 1 = (z + 1)(z^2 - z + 1)$

4. $125y^3 + 64 = (5y + 4)(25y^2 - 20y + 16)$

5. $27x^3 - 8 = (3x - 2)(9x^2 + 6x + 4)$

6. $f(x) = 2x^3 + 5x^2 + 3x - 8$. Possible rational zeros: $\pm 1, \pm 2, \pm 4, \pm 8, \pm \frac{1}{2}$.

7. $y = 4x^3 - 9x^2 + 3x + 20$. Possible rational zeros: $\pm 1, \pm 2, \pm 4, \pm 5, \pm 10, \pm 20, \pm \frac{1}{2}, \pm \frac{5}{2}, \pm \frac{1}{4}, \pm \frac{5}{4}$.

8. $g(x) = 6x^5 - 4x^3 - 5x - 4$. Possible rational zeros: $\pm 1, \pm 2, \pm 4, \pm \frac{1}{2}, \pm \frac{1}{3}, \pm \frac{2}{3}, \pm \frac{4}{3}, \pm \frac{1}{6}$.

9. $y = 6x^4 - 9x^3 + x + 14$. Possible rational zeros: $\pm 1, \pm 2, \pm 7, \pm 14, \pm \frac{1}{2}, \pm \frac{7}{2}, \pm \frac{1}{3}, \pm \frac{2}{3}, \pm \frac{7}{3}, \pm \frac{14}{3}, \pm \frac{1}{6}, \pm \frac{7}{6}$.

10. $f(x) = 9x^3 - 4x^2 + 2x - 15$. Possible rational zeros: $\pm 1, \pm 2, \pm 3, \pm 5, \pm 15, \pm \frac{1}{3}, \pm \frac{2}{3}, \pm \frac{5}{3}, \pm \frac{1}{9}, \pm \frac{2}{9}, \pm \frac{5}{9}$.

11. $f(x) = 2x^3 + 5x^2 + 3x - 8$, maximum of 1 positive real zero.
$f(-x) = -2x^3 + 5x^2 - 3x - 8$, maximum of 2 negative real zeros.

12. $g(x) = 4x^3 - 9x^2 + 3x + 20$, maximum of 2 positive real zeros.
$g(-x) = -4x^3 - 9x^2 - 3x + 20$, maximum of 1 negative real zero.

13. $f(x) = 6x^5 - 4x^3 - 5x - 4$, maximum of 1 positive real zero.
$f(-x) = -6x^5 + 4x^3 + 5x - 4$, maximum of 2 negative real zeros.

14. $g(x) = 6x^4 - 9x^3 + x^2 + 14$, maximum of 2 positive real zeros.
$g(-x) = 6x^4 + 9x^3 + x^2 + 14$, no negative real zeros.

15. $f(x) = 9x^3 - 4x^2 + 2x - 15$, maximum of 3 positive real zeros.
$f(-x) = -9x^3 - 4x^2 - 2x - 15$, no negative real zeros.

16.
$$\begin{array}{r} x^2-8 \\ x-3\overline{\smash{\big)}\ x^3-3x^2-8x+24} \\ \underline{-\left(x^3-3x^2\right)} \\ -8x+24 \\ \underline{-(-8x+24)} \end{array}$$

$x-3$ is a factor.

17.
$$\begin{array}{r} x^2-6x+9 \\ x+2\overline{\smash{\big)}\ x^3-4x^2-3x+12} \\ \underline{-\left(x^3+2x^2\right)} \\ -6x^2-3x \\ \underline{-\left(-6x^2-12x\right)} \\ 9x+12 \\ \underline{-(9x+18)} \\ -6 \end{array}$$

$x+2$ is not a factor.

18.
$$\begin{array}{r} 2x^2+9x+46 \\ x-6\overline{\smash{\big)}\ 2x^3-3x^2-8x+12} \\ \underline{-\left(2x^3-12x^2\right)} \\ 9x^2-8x \\ \underline{-\left(9x^2-54x\right)} \\ 46x+12 \\ \underline{-(46x-276)} \\ 288 \end{array}$$

$x-6$ is not a factor.

19.
$$\begin{array}{r} 2x+11 \\ x^2-3\overline{\smash{\big)}\ 2x^3+11x^2+12x-9} \\ \underline{-\left(2x^3-6x\right)} \\ 11x^2+18x-9 \\ \underline{-\left(11x^2-33\right)} \\ 18x+24 \end{array}$$

x^2-3 is not a factor.

20.
$$\begin{array}{r} x+3 \\ x^2-3x+9\overline{\smash{\big)}\ x^3+0x^2+0x+27} \\ \underline{-\left(x^3-3x^2+9x\right)} \\ 3x^2-9x+27 \\ \underline{-\left(3x^2-9x+27\right)} \\ 0 \end{array}$$

x^2-3x+9 is a factor.

21.
$$\begin{array}{r|rrrrr} 3 & 2 & 4 & -2 & 0 & 6 \\ & \downarrow & 6 & 30 & 84 & 252 \\ \hline & 2 & 10 & 28 & 84 & \underline{|258} \end{array}$$

$x=3$ is not a zero.

22.
$$\begin{array}{r|rrrr} -4 & 3 & -8 & -41 & 30 \\ & \downarrow & -12 & 80 & -156 \\ \hline & 3 & -20 & 39 & \underline{|-126} \end{array}$$

$x=-4$ is not a zero.

23.
$$\begin{array}{r|rrrrr} 2 & 5 & -3 & 2 & 0 & -3 \\ & \downarrow & 10 & 14 & 32 & 64 \\ \hline & 5 & 7 & 16 & 32 & \underline{|61} \end{array}$$

$x=2$ is not a zero.

24.
$$\begin{array}{r|rrrr} -1 & 3 & -8 & 5 & 16 \\ & \downarrow & -3 & 11 & -16 \\ \hline & 3 & -11 & 16 & \underline{|0} \end{array}$$

$x=-1$ is a zero.

25.
$$\begin{array}{r|rrrr} -2 & 1 & -7 & 2 & 40 \\ & & -2 & 18 & -40 \\ \hline & 1 & -9 & 20 & \underline{|0} \end{array}$$

$x = -2$ is a zero.

26.
$$\begin{array}{r|rrrr} 5 & 1 & -7 & 7 & +15 \\ & & 5 & -10 & -15 \\ \hline & 1 & -2 & -3 & \underline{|0} \end{array}$$

$x^3 - 7x^2 + 7x + 15 = (x-5)(x^2 - 2x - 3)$

$\qquad = (x-5)(x-3)(x+1)$

$\qquad x = 5 \quad x = 3 \quad x = -1$

27.
$$\begin{array}{r|rrrr} \frac{1}{2} & 2 & 3 & -18 & 8 \\ & & 1 & 2 & -8 \\ \hline & 2 & 4 & -16 & \underline{|0} \end{array}$$

$2x^3 + 3x^2 - 18x + 8 = \left(x - \frac{1}{2}\right)\left(2x^2 + 4x - 16\right)$

$= 2\left(x - \frac{1}{2}\right)\left(x^2 + 2x - 8\right) = 2\left(x - \frac{1}{2}\right)(x-2)(x+4)$

$x = \frac{1}{2} \quad x = 2 \quad x = -4$

28.
$$\begin{array}{r|rrrrr} -2 & 3 & 7 & -25 & -63 & -18 \\ & & -6 & -2 & 54 & 18 \\ \hline & 3 & 1 & -27 & -9 & \underline{|0} \end{array}$$

$3x^4 + 7x^3 - 25x^2 - 63x - 18 = (x+2)\left(3x^3 + x^2 - 27x - 9\right)$

$$\begin{array}{r|rrrrr} -2 & 3 & 7 & -25 & -63 & -18 \\ & & -6 & -2 & 54 & 18 \\ \hline 3 & 3 & 1 & -27 & -9 & \underline{|0} \\ & & 9 & 30 & 9 & \\ \hline & 3 & 10 & 3 & \underline{|0} \end{array}$$

$(x+2)\left(3x^3 + x^2 - 27x - 9\right) = (x+2)(x-3)\left(3x^2 + 10x + 3\right)$

$= (x+2)(x-3)(x+3)(3x+1)$

$x = -2 \quad x = 3 \quad x = -3 \quad x = -\frac{1}{3}$

29.
$$\begin{array}{r|rrrrr} 2i & 1 & -5 & 10 & -20 & 24 \\ & & 2i & -4-10i & 20+12i & -24 \\ \hline -2i & 1 & -5+2i & 6-10i & 12i & \underline{|0} \\ & & -2i & 10i & -12i & \\ \hline & 1 & -5 & 6 & \underline{|0} \end{array}$$

$x^4 - 5x^3 + 10x^2 - 20x + 24 = (x-2i)(x+2i)\left(x^2 - 5x + 6\right)$

$= (x-2i)(x+2i)(x-2)(x-3)$

$x = 2i \quad x = -2i \quad x = 2 \quad x = 3$

30.
$$\begin{array}{r|rrrrr} -3i & 1 & 0 & 13 & 0 & 36 \\ & & -3i & -9 & -12i & -36 \\ \hline 3i & 1 & -3i & 4 & -12i & \underline{|0} \\ & & 3i & 0 & 12i & \\ \hline & 1 & 0 & 4 & \underline{|0} \end{array}$$

$x^4 + 13x^2 + 36 = (x+3i)(x-3i)\left(x^2 + 4\right)$

$= (x+3i)(x-3i)(x+2i)(x-2i)$

$x = -3i \quad x = 3i \quad x = -2i \quad x = 2i$

31. $x^3 - x^2 - 17x - 15 \geq 0$

$(x-5)(x+3)(x+1) \geq 0$

$-3 \leq x \leq -1 \quad x \geq 5$

Interval	Value	$x-5$	$x+3$	$x+1$	$x^3 - x^2 - 17x - 15$
$x < -3$	-4	$-$	$-$	$-$	$-$
$-3 < x < -1$	-2	$-$	$+$	$-$	$+$
$-1 < x < 5$	0	$-$	$+$	$+$	$-$
$x > 5$	6	$+$	$+$	$+$	$+$

32. $2x^3 + 3x^2 - 18x + 8 < 0$

$(x-2)(x+4)(2x-1) < 0$

$x < -4 \quad \frac{1}{2} < x < 2$

Interval	Value	$x-2$	$x+4$	$2x-1$	$2x^3 + 3x^2 - 18x + 8$
$x < -4$	-5	$-$	$-$	$-$	$-$
$-4 < x < \frac{1}{2}$	0	$-$	$+$	$-$	$+$
$\frac{1}{2} < x < 2$	1	$-$	$+$	$+$	$-$
$x > 2$	3	$+$	$+$	$+$	$+$

33. $x^4 - 5x^2 + 6 \le 0$

$(x^2 - 3)(x^2 - 2) \le 0$

$(x + \sqrt{3})(x - \sqrt{3})(x + \sqrt{2})(x - \sqrt{2}) \le 0$

$-\sqrt{3} \le x \le -\sqrt{2}$ $\sqrt{2} \le x \le \sqrt{3}$

Interval	Value	$x + \sqrt{3}$	$x - \sqrt{3}$	$x + \sqrt{2}$	$x - \sqrt{2}$	$x^4 - 5x^2 + 6$
$x < -\sqrt{3}$	−2	−	−	−	−	+
$-\sqrt{3} < x < -\sqrt{2}$	−1.5	+	−	−	−	−
$-\sqrt{2} < x < \sqrt{2}$	0	+	−	+	−	+
$\sqrt{2} < x < \sqrt{3}$	1.5	+	−	+	+	−
$x > \sqrt{3}$	2	+	+	+	+	+

34. $4x^5 - 32x^4 + 48x^3 > 0$

$4x^3(x^2 - 8x + 12) > 0$

$4x^3(x - 2)(x - 6) > 0$

$0 < x < 2$ $x > 6$

Interval	Value	$4x^3$	$x - 2$	$x - 6$	$4x^5 - 32x^4 + 48x^3$
$x < 0$	−1	−	−	−	−
$0 < x < 2$	1	+	−	−	+
$2 < x < 6$	3	+	+	−	−
$x > 6$	7	+	+	+	+

35. $x^4 - 4x^2 - 45 \le 0$

$(x^2 - 9)(x^2 + 5) \le 0$

$(x + 3)(x - 3)(x^2 + 5) \le 0$

$-3 \le x \le 3$

Interval	Value	$x + 3$	$x - 3$	$x^2 + 5$	$x^4 - 4x^2 - 45$
$x < -3$	−4	−	−	+	+
$-3 < x < 3$	0	+	−	+	−
$x > 3$	4	+	+	+	+

36. $y = x^3 - 7x^2 + 7x - 15$

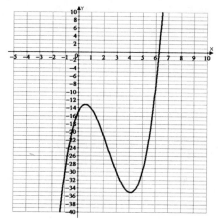

37. $y = 2x^3 + 3x^2 - 18x + 8$

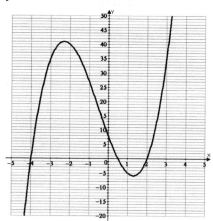

38. $y = x^4 - 5x^2 + 6$

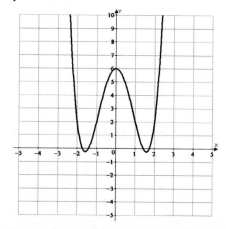

39. $y = 2x^4 - 16x^3 + 24x^2$

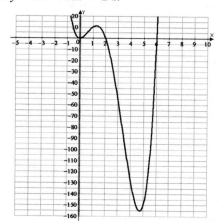

40. $y = x^4 - 4x^2 - 45$

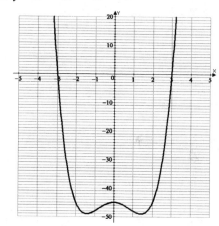

Chapter 11

1. Domain of $\frac{7x}{21x^2}$ is $x \neq 0$, $\frac{7x}{21x^2} = \frac{1}{3x}$.

2. Domain of $\frac{5y+25}{y^2-25}$ is $y \neq \pm 5$,

   $\frac{5y+25}{y^2-25} = \frac{5(y+5)}{(y+5)(y-5)} = \frac{5}{y-5}$.

3. Domain of $\frac{x^2-8x+15}{x^2-7x+10}$ is $x \neq 2$

   and $x \neq 5$, $\frac{x^2-8x+15}{x^2-7x+10} = \frac{(x-3)(x-5)}{(x-2)(x-5)} = \frac{x-3}{x-2}$.

4. Domain of $\frac{x^2-x-42}{x^2-49}$ is $x \neq \pm 7$,

   $\frac{x^2-x-42}{x^2-49} = \frac{(x+6)(x-7)}{(x+7)(x-7)} = \frac{x+6}{x+7}$.

5. Domain of $\dfrac{2y^2+7y+6}{2y^2+y-3} >$ is $y \ne 1$ and $y \ne -\dfrac{3}{2}$, $\dfrac{2y^2+7y+6}{2y^2+y-3} = \dfrac{\cancel{(2y+3)}(y+2)}{\cancel{(2y+3)}(y-1)} = \dfrac{y+2}{y-1}$.

6. $\dfrac{3x+2}{x} + \dfrac{5x+7}{2x} = \dfrac{2(3x+2)}{2x} + \dfrac{5x+7}{2x} = \dfrac{6x+4+5x+7}{2x} = \dfrac{11x+11}{2x}$

7. $\dfrac{a}{a+3} + \dfrac{2a}{3a+9} = \dfrac{3a}{3(a+3)} + \dfrac{2a}{3a+9} = \dfrac{5a}{3a+9}$

8. $\dfrac{2}{x} + \dfrac{3}{x+1} = \dfrac{2(x+1)}{x(x+1)} + \dfrac{3x}{x(x+1)} = \dfrac{2x+2+3x}{x(x+1)} = \dfrac{5x+2}{x(x+1)}$

9. $\dfrac{x+2}{x+3} - \dfrac{x+3}{x+4} = \dfrac{(x+2)(x+4)}{(x+3)(x+4)} - \dfrac{(x+3)(x+3)}{(x+4)(x+3)} = \dfrac{x^2+6x+8}{(x+4)(x+3)} - \dfrac{x^2+6x+9}{(x+4)(x+3)} = \dfrac{-1}{(x+4)(x+3)}$

10. $\dfrac{y-2}{y^2-5y+4} - \dfrac{y-2}{y^2-16} = \dfrac{y-2}{(y-4)(y-1)} - \dfrac{y-2}{(y+4)(y-4)} = \dfrac{(y-2)(y+4)}{(y-4)(y-1)(y+4)} - \dfrac{(y-2)(y-1)}{(y+4)(y-4)(y-1)}$

$= \dfrac{y^2+2y-8}{(y-4)(y-1)(y+4)} - \dfrac{y^2-3y+2}{(y+4)(y-4)(y-1)} = \dfrac{5y-10}{(y+4)(y-4)(y-1)}$

11. $\dfrac{x^2-1}{6-3x} \cdot \dfrac{12-6x}{2x+2} = \dfrac{(x+1)(x-1)}{\cancel{3}\cancel{(x-2)}} \cdot \dfrac{\cancel{6}\,\cancel{(x-2)}}{\cancel{2}\,\cancel{(x+1)}} = x-1$

12. $\dfrac{x^2+7x+10}{x^2-2x+1} \cdot \dfrac{x^2-6x+5}{x^2-25} = \dfrac{(x+2)\cancel{(x+5)}}{(x-1)\cancel{(x-1)}} \cdot \dfrac{\cancel{(x-1)}\cancel{(x-5)}}{\cancel{(x+5)}\cancel{(x-5)}} = \dfrac{x+2}{x-1}$

13. $\dfrac{x^2-16x+28}{2x-28} \cdot \dfrac{3x^3}{6x^2-12x} = \dfrac{\cancel{(x-2)}\cancel{(x-14)}}{2\cancel{(x-14)}} \cdot \dfrac{3x^3}{6x\cancel{(x-2)}} = \dfrac{3\cdot x\cdot x\cdot \cancel{x}}{2\cdot 6\cdot \cancel{x}} = \dfrac{3x^2}{12} = \dfrac{x^2}{4}$

14. $\dfrac{x^2-5x+6}{x^2-7x+10} \cdot \dfrac{x^2-4x-5}{x^2-2x-3} = \dfrac{\cancel{(x-2)}\cancel{(x-3)}}{\cancel{(x-2)}\cancel{(x-5)}} \cdot \dfrac{\cancel{(x-5)}\cancel{(x+1)}}{\cancel{(x-3)}\cancel{(x+1)}} = 1$

15. $\dfrac{x^2+x-42}{x^2-36} \cdot \dfrac{x^2-49}{x^2-x-42} = \dfrac{(x+7)\cancel{(x-6)}}{(x+6)\cancel{(x-6)}} \cdot \dfrac{(x+7)\cancel{(x-7)}}{\cancel{(x-7)}(x+6)} = \dfrac{(x+7)^2}{(x+6)^2}$

16. $\dfrac{5x^2-20}{6} \div \dfrac{x^2-4x+4}{3} = \dfrac{5x^2-20}{6} \cdot \dfrac{3}{x^2-4x+4} = \dfrac{5(x+2)\cancel{(x-2)}}{\cancel{6}\,_{2}} \cdot \dfrac{\cancel{3}}{(x-2)\cancel{(x-2)}} = \dfrac{5(x+2)}{2(x-2)}$

17. $\dfrac{5x^2-20}{x^2-5x+6} \div \dfrac{6x+12}{3x-6} = \dfrac{5x^2-20}{x^2-5x+6} \cdot \dfrac{3x-6}{6x+12} = \dfrac{5\cancel{(x+2)}(x-2)}{\cancel{(x-2)}(x-3)} \cdot \dfrac{3\cancel{(x-2)}}{6\cancel{(x+2)}} = \dfrac{5(x-2)}{2(x-3)}$

18. $\dfrac{2x^2-x}{4x^2-1} \div \dfrac{4x}{8x+4} = \dfrac{2x^2-x}{4x^2-1} \cdot \dfrac{8x+4}{4x} = \dfrac{\cancel{x}\cancel{(2x-1)}}{\cancel{(2x+1)}\cancel{(2x-1)}} \cdot \dfrac{4\cancel{(2x+1)}}{4\cancel{x}} = 1$

19. $\dfrac{1-16y^2}{y^4-16} \div \dfrac{y-4y^2}{y^2-4y+4} = \dfrac{1-16y^2}{y^4-16} \cdot \dfrac{y^2-4y+4}{y-4y^2} = \dfrac{(1+4y)\cancel{(1-4y)}}{(y^2+4)(y+2)\cancel{(y-2)}} \cdot \dfrac{\cancel{(y-2)}(y-2)}{y\cancel{(1-4y)}} = \dfrac{(1+4y)(y-2)}{y(y+2)(y^2+4)}$

20. $\dfrac{a^2+4a-45}{4a-20} \div \dfrac{a^2-81}{3a-3} = \dfrac{a^2+4a-45}{4a-20} \cdot \dfrac{3a-3}{a^2-81} = \dfrac{\cancel{(a+9)}\cancel{(a-5)}}{4\cancel{(a-5)}} \cdot \dfrac{3(a-1)}{\cancel{(a+9)}(a-9)} = \dfrac{3(a-1)}{4(a-9)}$

21. $\dfrac{4+\frac{2}{x}}{\frac{4}{x}-2} \cdot \dfrac{x}{x} = \dfrac{4x+2}{4-2x} = \dfrac{\cancel{2}(2x+1)}{\cancel{2}(2-x)} = \dfrac{2x+1}{2-x}$

22. $\dfrac{\frac{2}{3x}-\frac{3}{2x}}{\frac{5}{2x}+\frac{2}{3x}}\cdot\dfrac{6x}{6x}=\dfrac{2\cdot2-3\cdot3}{5\cdot3+2\cdot2}=\dfrac{4-9}{15+4}=\dfrac{-5}{19}$

23. $\dfrac{\frac{1}{x}-1}{1+\frac{1}{x}}\cdot\dfrac{x}{x}=\dfrac{1-x}{x+1}$

24. $\dfrac{\frac{3}{x-1}+5}{\frac{1}{x-5}+3}\cdot\dfrac{(x-1)(x-5)}{(x-1)(x-5)}=\dfrac{3(x-5)+5(x-1)(x-5)}{x-1+3(x-1)(x-5)}=\dfrac{3x-15+5(x^2-6x+5)}{x-1+3(x^2-6x+5)}=\dfrac{3x-15+5x^2-30x+25}{x-1+3x^2-18x+15}=\dfrac{5x^2-27x+10}{3x^2-17x+14}=\dfrac{(5x-2)(x-5)}{(3x-14)(x-1)}$

25. $\dfrac{\frac{1}{x-3}+\frac{x}{x+2}}{\frac{x}{x-3}-\frac{1}{x+2}}\cdot\dfrac{(x-3)(x+2)}{(x-3)(x+2)}=\dfrac{x+2+x(x-3)}{x(x+2)-(x-3)}=\dfrac{x^2-2x+2}{x^2+x+3}$

26. $\dfrac{4}{t-6}=\dfrac{3}{t}\to 4t=3t-18\to t=-18$

27. $\dfrac{x-3}{x+7}=\dfrac{x-2}{x+4}\to x^2+x-12=x^2+5x-14\to x-12=5x-14\to 2=4x\to x=\frac{1}{2}$

28. $\dfrac{3}{x-4}=\dfrac{x+4}{3}\Rightarrow x^2-16=9\Rightarrow x^2=25\Rightarrow x=\pm5$

29. $\dfrac{4}{x+1}=\dfrac{x}{5}\to x^2+x=20\to x^2+x-20=0\to(x+5)(x-4)=0\to x=-5, x=4$

30. $\dfrac{3x+1}{x-1}=\dfrac{x+2}{x-2}\to 3x^2-5x-2=x^2+x-2\to 2x^2-6x=0\to 2x(x-3)=0\to x=0, x=3$

31. $12\left(\dfrac{2x-5}{6}-\dfrac{1}{4}\right)=12\left(\dfrac{3-x}{12}\right)$

    $2(2x-5)-3=3-x$

    $4x-13=3-x$

    $5x=16$

    $x=3.2$

32. $6a\left(\dfrac{3}{a}-\dfrac{1}{2a}+\dfrac{1}{3a}\right)=6\cdot6a$

    $18-3+2=36a$

    $17=36a$

    $a=\dfrac{17}{36}$

33. $6x(x-1)\left(\dfrac{1}{x}+\dfrac{1}{x-1}\right)=\dfrac{5}{6}\cdot6x(x-1)$

    $6(x-1)+6x=5x(x-1)$

    $6x-6+6x=5x^2-5x$

    $5x^2-17x+6=0$

    $(5x-2)(x-3)=0$

    $x=\dfrac{2}{5}\qquad x=3$

34. $(x+2)(x-2)\left(\dfrac{7}{x+2}\right)=(x+2)(x-2)\left(2-\dfrac{3}{x-2}\right)$

    $7(x-2)=2(x+2)(x-2)-3(x+2)$

    $7x-14=2x^2-8-3x-6$

    $7x-14=2x^2-3x-14$

    $0=2x^2-10x$

    $2x(x-5)=0$

    $2x=0\qquad x-5=0$

    $x=0\qquad x=5$

35. $x(x-1)(x+1)\left(\dfrac{6}{x-1}+\dfrac{8}{x}\right)=x(x-1)\dfrac{5}{x(x+1)}$

    $6x(x+1)+8(x-1)(x+1)=5(x-1)$

    $6x^2+6x+8x^2-8=5x-5$

    $14x^2+x-3=0$

    $(2x+1)(7x-3)=0$

    $2x+1=0\qquad 7x-3=0$

    $x=-\dfrac{1}{2}\qquad x=\dfrac{3}{7}$

36. Domain of $f(x) = \frac{2}{x-3} + 1$ is $x \neq 3$.
    Vertical asymptote: $x = 3$.
    Horizontal asymptote: $y = 1$.

37. Domain of $g(x) = \frac{2x-3}{x+1}$ is $x \neq -1$.
    Vertical asymptote: $x = -1$.
    Horizontal asymptote: $y = 2$.

38. Domain of $f(x) = \frac{x}{x^2-4}$ is $x \neq \pm 2$.
    Vertical asymptotes: $x = 2$, $x = -2$.
    Horizontal asymptote: $y = 0$.

39. Domain of $g(x) = \frac{x^2-1}{x+4}$ is $x \neq -4$.
    Vertical asymptote: $x = -4$.
    Oblique asymptote: $y = x - 4$.

40. Domain of $f(x) = \frac{2x-5}{x+2}$ is $x \neq -2$.
    Vertical asymptote: $x = -2$.
    Horizontal asymptote: $y = 2$.

41. $f(x) = \frac{2}{x-3} + 1$

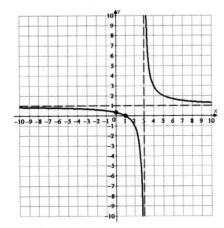

42. $g(x) = \frac{2x-3}{x+1}$

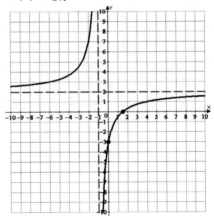

43. $f(x) = \frac{2x-5}{x+2}$

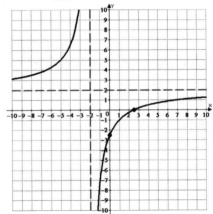

44. $g(x) = \frac{x^2-1}{x+4}$

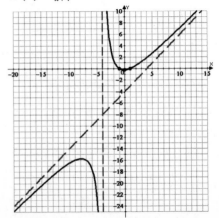

45. $f(x) = \frac{x}{x^2 - 4}$

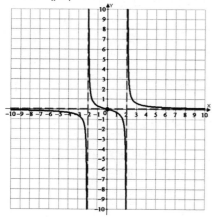

Chapter 12

1. $2^7 = 128$ is equivalent to $\log_2 128 = 7$

2. $x^y = z$ is equivalent to $\log_x z = y$

3. $\log_3 81 = 4$ is equivalent to $3^4 = 81$

4. $\log_r t = v$ is equivalent to $r^v = t$

5. $\log_5 144 = \frac{\log 144}{\log 5} = \frac{2.158362492\ldots}{0.6989700043\ldots} \approx 3.087918621$

   or $\log_5 144 = \frac{\ln 144}{\ln 5} = \frac{4.9698133\ldots}{1.609437912\ldots} \approx 3.087918621$

6. $\log_7(3y^5) = \log_7(3) + 5\log_7(y)$

7. $\log_2(8n^4) = \log_2(8) + 4\log_2(n) = 3 + 4\log_2(n)$

8. $\log_b\left(\frac{4x^2}{y}\right) = \log_b\left(4x^2\right) - \log_b(y) = \log_b(4) + 2\log_b(x) - \log_b(y)$

9. $\log_3\left(x^2\sqrt{y}\right) = \log_3\left(x^2\right) + \log_3\left(y^{\frac{1}{2}}\right) = 2\log_3(x) + \frac{1}{2}\log_3(y)$

10. $\log\left(\frac{x^2-4}{x^2+3x-4}\right) = \log\left(x^2-4\right) - \log\left(x^2+3x-4\right)$

    $= \log\left[(x+2)(x-2)\right] - \log\left[(x+4)(x-1)\right]$

    $= \log(x+2) + \log(x-2) - \log(x+4) - \log(x-1)$

11. $3\log x + \frac{1}{2}\log y = \log x^3 + \log\sqrt{y} = \log\left(x^3\sqrt{y}\right)$

12. $8\log_3 t - 5\log_3 u = \log_3 t^8 - \log_3 u^5 = \log_3\left(\frac{t^8}{u^5}\right)$

13. $\frac{1}{2}\log(x+4)+3\log(y+2)=\log\left(\sqrt{x+4}\right)+\log(y+2)^3=\log\left[(y+2)^3\sqrt{x+4}\right]$

14. $\ln(x-5)+\ln(x+5)-2\ln(x-1)=\ln\left(x^2-25\right)-\ln(x-1)^2=\ln\left[\frac{x^2-25}{(x-1)^2}\right]$

15. $4\log_2 a+\frac{1}{2}\log_2 b-3\log_2 c=\log_2 a^4\sqrt{b}-\log_2 c^3=\log_2\left(\frac{a^4\sqrt{b}}{c^3}\right)$

16. $2^{1-2x}=4^{x+3}$

$2^{1-2x}=\left(2^2\right)^{x+}$

$2^{1-2x}=2^{2x+6}$

$1-2x=2x+6$

$-5=4x$

$\frac{-5}{4}=x$

17. $81^{5x-7}=27^{7x-11}$

$\left(3^4\right)^{5x-7}=\left(3^3\right)^{7x-11}$

$3^{20x-28}=3^{21x-33}$

$20x-28=21x-33$

$5=x$

18. $\log\left(6^x\right)=\log\left(5^{2x-1}\right)$

$x\log 6=(2x-1)\log 5$

$x\log 6=2x\log 5-\log 5$

$\log 5=2x\log 5-x\log 6$

$\log 5=x\left(2\log 5-\log 6\right)$

$x=\frac{\log 5}{2\log 5-\log 6}\approx 1.128$

19. $\log\left[2\left(5^{3x}\right)\right]=\log(41)$

$\log 2+\log\left(5^{3x}\right)=\log 41$

$\log 2+3x\log 5=\log 41$

$3x\log 5=\log 41-\log 2$

$x=\frac{\log 41-\log 2}{3\log 5}\approx 0.626$

20. $\ln\left(1,500e^{0.06t}\right)=\ln(1,850)$

$\ln(1,500)+0.06t\ln e=\ln(1,850)$

$0.06t=\ln 1,850-\ln 1,500$

$t=\frac{\ln 1,850-\ln 1,500}{0.06}\approx 3.495$

21. $\left(3^x\right)^2-5\left(3^x\right)+6=0$

$a^2-5a+6=0$

$(a-2)(a-3)=0$

$a=2$ $a=3$

$3^x=2$ $3^x=3$

$x\log 3=\log 2$ $x=1$

$x=\frac{\log 2}{\log 3}\approx 0.631$

22. $\left(5^{2x}\right)^2-9\left(5^{2x}\right)+20=0$

$a^2-9a+20=0$

$(a-4)(a-5)=0$

$a=4$ $a=5$

$5^{2x}=4$ $5^{2x}=5$

$2x\log 5=\log 4$ $2x=1$

$x=\frac{\log 4}{2\log 5}\approx 0.431$ $x=\frac{1}{2}$

23. $\left(\frac{1}{2}\right)^{2x} - \left(\frac{1}{2}\right)^{x} - 30 = 0$

$a^2 - a - 30 = 0$

$(a-6)(a+5) = 0$

$a = 6$

$\left(\frac{1}{2}\right)^{x} = 6$

$x\log\left(\frac{1}{2}\right) = \log(6)$

$x = \frac{\log 6}{\log\left(\frac{1}{2}\right)} \approx -2.585$

$a = -5$

$\left(\frac{1}{2}\right)^{x} = -5$

reject

24. $e^{2x} - 4\left(e^{x}\right) + 4 = 0$

$a^2 - 4a + 4 = 0$

$(a-2)^2 = 0$

$a = 2$

$e^{x} = 2$

$x = \ln 2 \approx 0.693$

25. $9^{x} - 12\left(3^{x}\right) + 27 = 0$

$3^{2x} - 12\left(3^{x}\right) + 27 = 0$

$a^2 - 12a + 27 = 0$

$(a-9)(a-3) = 0$

$a = 9 \qquad a = 3$

$3^{x} = 9 \qquad 3^{x} = 3$

$x = 2 \qquad x = 1$

26. $\log_5(2x - 5) = \log_5(x + 7)$

$2x - 5 = x + 7$

$x = 12$

27. $\ln\left(x^2 - 8x + 5\right) = \ln(4 - 6x)$

$x^2 - 8x + 5 = 4 - 6x$

$x^2 - 2x + 1 = 0$

$(x-1)^2 = 0$

$x = 1$

reject – no solution

28. $\log_2(6x + 5) = 5$

$6x + 5 = 2^5$

$6x + 5 = 32$

$6x = 27$

$x = \frac{27}{6} = \frac{9}{2}$

29. $\log_4(x) + \log_4(x - 3) = 1$

$\log_4\left(x^2 - 3x\right) = 1$

$x^2 - 3x = 4^1$

$x^2 - 3x - 4 = 0$

$(x-4)(x+1) = 0$

$x = -1$

$x = 4$

reject

30. $\log(2x - 3) - \log(x + 1) = \log(x - 1)$

$\log\left(\frac{2x-3}{x+1}\right) = \log(x - 1)$

$\frac{2x-3}{x+1} = x - 1$

$2x - 3 = x^2 - 1$

$x^2 - 2x + 2 = 0$

no real solutions

31. $f(x) = 2^{x-3} + 5$

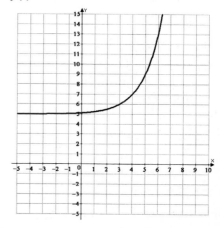

32. $g(x) = -\frac{1}{2}\left(3^{x+4}\right) - 2$

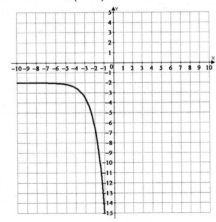

35. $f(x) = \ln(4 - x) + 3$

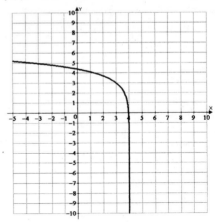

33. $f(x) = 3\log_2(x - 1)$

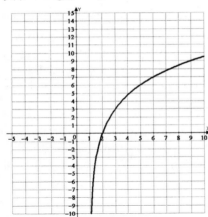

36. You must deposit approximately \$4,885.

$$A = P\left(1 + \frac{r}{n}\right)^{nt}$$

$$10,000 = P\left(1 + \frac{.04}{4}\right)^{4 \cdot 18}$$

$$10,000 = P(1.01)^{72}$$

$$P = \frac{10,000}{(1.01)^{72}} \approx 4,884.96$$

37. It will take approximately 21 years and 8 months.

$$A = Pe^{rt}$$

$$2 = 1 \cdot e^{0.032t}$$

$$\ln 2 = 0.032t \ln e$$

$$t = \frac{\ln 2}{0.032} \approx 21.66$$

34. $g(x) = 5 - \log_3(x + 2)$

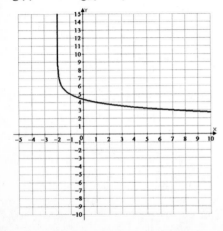

38. The growth rate is approximately 0.0335 bacteria per hour. The population of the colony will reach 1,000,000 after approximately 275 hours, or about 11.5 days.

$$A = Pe^{rt}$$
$$500 = 100 \cdot e^{r \cdot 48}$$
$$5 = e^{48r}$$
$$\ln 5 = 48r$$
$$r = \frac{\ln 5}{48} \approx 0.0335$$

$$A = Pe^{rt}$$
$$1,000,000 = 100 \cdot e^{0.0335t}$$
$$10,000 = e^{0.0335t}$$
$$\ln(10,000) = 0.0335t$$
$$t = \frac{\ln(10,000)}{0.0335} \approx 274.69$$

39. The rate of decay is approximately $-1.057 \cdot 10^{-4}$. (A negative rate indicates decay.) There will be approximately 9.963 grams left in 2050 from 10 grams measured in 2015.

$$A = Pe^{rt}$$
$$\tfrac{1}{2} = 1 \cdot e^{6560r}$$
$$\ln\left(\tfrac{1}{2}\right) = 6560r$$
$$r = \frac{\ln\left(\tfrac{1}{2}\right)}{6560} \approx -1.057 \times 10^{-4}$$

$$A = Pe^{rt}$$
$$A \approx 10 \cdot e^{\left(-1.057 \times 10^{-4}\right) \cdot 35}$$
$$A \approx 10 \cdot e^{-0.0037}$$
$$A \approx 10(0.996308\ldots)$$
$$A \approx 9.963$$

40. The concentration of hydrogen atoms in lemon juice is approximately $H^+ = 10^{-2.1} \approx 0.0079$.

$$pH = -\log\left(H^+\right)$$
$$2.1 = -\log\left(H^+\right)$$
$$-2.1 = \log\left(H^+\right)$$
$$H^+ = 10^{-2.1} \approx 0.0079$$

Chapter 13

1. $10 \cdot 10 \cdot 10 = 1,000$

2. $10 \cdot 9 \cdot 8 = 720$

3. $36 \cdot 36 \cdot 36 \cdot 36 \cdot 36 \cdot 36 = 2,176,782,336$. (In some states with similar patterns, the pattern must be one group of three letters and one of three numbers. In that case, the number is $26 \cdot 26 \cdot 26 \cdot 10 \cdot 10 \cdot 10 = 17,576,000$.)

4. There are 10 choices for each digit, so there are $10 \cdot 10 \cdot 10$ different area codes, $10 \cdot 10 \cdot 10$ different exchanges, and $10 \cdot 10 \cdot 10 \cdot 10$ different numbers: $10^3 \cdot 10^3 \cdot 10^4 = 10,000,000,000$.

5. $59 \cdot 58 \cdot 57 \cdot 56 \cdot 55 \cdot 35 \approx 2.1 \cdot 10^{10}$

6. $_4P_3 = \frac{4!}{(4-3)!} = 24$

7. $_6P_2 = \frac{6!}{(6-2)!} = 30$

8. $_{10}C_4 = \frac{10!}{(10-4)!4!} = \frac{10 \cdot 9 \cdot 8 \cdot 7}{4 \cdot 3 \cdot 2 \cdot 1} = 210$

9. $_7C_3 = \frac{7!}{(7-3)!3!} = \frac{7 \cdot 6 \cdot 5}{3 \cdot 2 \cdot 1} = 35$

10. $_{14}C_5 = \frac{14!}{(14-5)!5!} = \frac{14\cdot13\cdot12\cdot11\cdot10}{5\cdot4\cdot3\cdot2\cdot1} = 2{,}002$

```
                          1
                      1        1
                  1       2        1
              1       3      [3]       1
                            #12
          1       4      6       4      [1]
                                       #14
      1       5      10     10     [5]      1
                                  #11
  1       6      15    [20]    15      6      1
                       #13
1     7      21     35     35     21     7     1
1    [8]     28     56     70     56     28     8     1
     #15
```

11. $_5C_4 = 5$

12. $_3C_2 = 3$

13. $_6C_3 = 20$

14. $_4C_4 = 1$

15. $_8C_1 = 8$

16. P(head on penny and tail on dime) $= P$(head on penny) $\cdot P$(tail on dime) $= \frac{1}{2}\cdot\frac{1}{2} = \frac{1}{4}$

17. P(red or orange) $= P$(red) $+ P$(orange) $- P$(red and orange) $= \frac{4}{15}+\frac{3}{15}-\frac{0}{15} = \frac{7}{15}$

18. P(2nd blue|1st orange) $= \frac{P(\text{orange then blue})}{P(\text{orange})} = \frac{\frac{3\cdot7}{15\cdot14}}{\frac{3}{15}} = \frac{1}{2}$

19. P(blue then yellow) $= \frac{7}{15}\cdot\frac{1}{14} = \frac{1}{30}$

20. P(contract the disease and require hospitalization) $= 0.45 \cdot 0.10 = 0.045$. There is a 4.5% probability that you will contract the virus and require hospitalization.

21. $(x-2)^3 = \binom{3}{0}x^3(-2)^0 + \binom{3}{1}x^2(-2)^1 + \binom{3}{2}x^1(-2)^2 + \binom{3}{3}x^0(-2)^3$

$\qquad = 1x^3 + 3(-2)x^2 + 3(4)x + 1(-8)$

$\qquad = x^3 - 6x^2 + 12x - 8$

22. $(x+3)^4 = \binom{4}{0}x^4 \cdot 3^0 + \binom{4}{1}x^3 \cdot 3^1 + \binom{4}{2}x^2 \cdot 3^2 + \binom{4}{3}x^1 \cdot 3^3 + \binom{4}{4}x^0 \cdot 3^4$

$\qquad = 1x^4 + 4 \cdot 3x^3 + 6 \cdot 9x^2 + 4 \cdot 27x + 1 \cdot 81$

$\qquad = x^4 + 12x^3 + 54x^2 + 108x + 81$

23. $(2x+1)^5 = \binom{5}{0}(2x)^5(1)^0 + \binom{5}{1}(2x)^4(1)^1 + \binom{5}{2}(2x)^3(1)^2 + \binom{5}{3}(2x)^2(1)^3$

$\qquad\qquad + \binom{5}{4}(2x)^1(1)^4 + \binom{5}{5}(2x)^0(1)^5$

$\qquad = 1\left(32x^5\right)(1) + 5\left(16x^4\right)(1) + 10\left(8x^3\right)(1) + 10\left(4x^2\right)(1) + 5(2x)(1) + 1(1)$

$\qquad = 32x^5 + 80x^4 + 80x^3 + 40x^2 + 10x + 1$

24. $(3x-2)^4 = \binom{4}{0}(3x)^4(-2)^0 + \binom{4}{1}(3x)^3(-2)^1 + \binom{4}{2}(3x)^2(-2)^2$

$\qquad\qquad + \binom{4}{3}(3x)^1(-2)^3 + \binom{4}{4}(3x)^0(-2)^4$

$\qquad = 1\left(81x^4\right)(1) + 4\left(27x^3\right)(-2) + 6\left(9x^2\right)(4) + 4(3x)(-8) + 1(16)$

$\qquad = 81x^4 - 216x^3 + 216x^2 - 96x + 16$

25. $(x+1)^7 = \binom{7}{0}x^7 \cdot 1^0 + \binom{7}{1}x^6 \cdot 1^1 + \binom{7}{2}x^5 \cdot 1^2 + \binom{7}{3}x^4 \cdot 1^3 + \binom{7}{4}x^3 \cdot 1^4$

$\qquad\qquad + \binom{7}{5}x^2 \cdot 1^5 + \binom{7}{6}x^1 \cdot 1^6 + \binom{7}{7}x^0 \cdot 1^7$

$\qquad = 1x^7 + 7x^6 + 21x^5 + 35x^4 + 35x^3 + 21x^2 + 7x + 1$

Extra Practice Answers

Chapter 1

1. 6.725: rational, real

2. −92: integer, rational, real

3. $\sqrt{182}$: irrational, real

4. −5: integer, rational, real

5. $\frac{42}{5}$: rational, real

6. $9(2 − 5) \div 3 + 7 − 12 = 9(−3) \div 3 + 7 − 12 =$
   $−27 \div 3 + 7 − 12 = −9 + 7 − 12 = −2 − 12 =$
   $−14$

7. $5a − 12b \div 4 − 8b + 12a =$
   $5a − 3b − 8b + 12a =$
   $17a − 11b$

8. $15(2x + 4) − 25x + 3(1 − 10y) =$
   $30x + 60 − 25x + 3 − 30y =$
   $5x − 30y + 63$

9. $\frac{18x}{−2x} − \frac{24x+81}{3} = −9 − \left(8x + 27\right) = −9 − 8x − 27 = −8x − 36$

10. $\frac{6\left(12t^2 −15t\right)}{9t} = \frac{72t^2 −90t}{9t} = 8t − 10$

11. $5x + 3(4x + 9) = 13 − 8(4 − 5x)$
    $5x + 12x + 27 = 13 − 32 + 40x$
    $17x + 27 = −19 + 40x$
    $46 = 23x$
    $2 = x$

12. $12 − (17x + 15) = 27 − 13x$
    $12 − 17x − 15 = 27 − 13x$
    $−3 − 17x = 27 − 13x$
    $−30 = 4x$
    $−7.5 = x$

13. $5(6x + 3) = 8 − (4x + 15) + 2x$
    $30x + 15 = 8 − 4x − 15 + 2x$
    $30x + 15 = −2x − 7$
    $32x = −22$
    $x = \frac{−22}{32} = −\frac{11}{16}$

14. $1 − 4(x − 2) < 2 + 2x − 9$
    $1 − 4x + 8 < 2 + 2x − 9$
    $−4x + 9 < 2x − 7$
    $−6x < −16$
    $x > \frac{−16}{−6}$
    $x > \frac{8}{3}$

15. $3x - (12 - 5x) \geq 9 + 6x - 21$
$3x - 12 + 5x \geq 9 + 6x - 21$
$8x - 12 \geq 6x - 12$
$2x \geq 0$
$x \geq 0$

16. $y = -\frac{2}{3}x + 5$

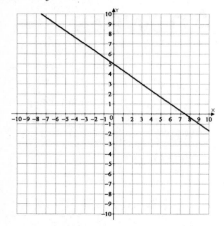

17. $y = 4x - 7$

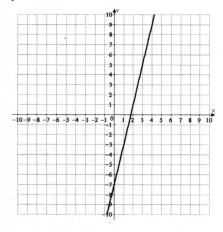

18. $3x - 2y = 12$

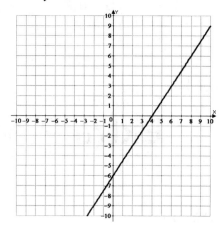

19. $7x + 3y = -21$

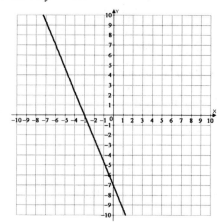

20. $4y - 22 = 11x$

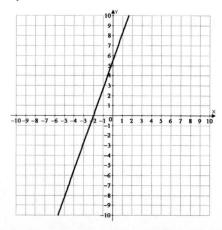

21. $y = \frac{2}{5}x - 3$

22. $y - 5 = 4(x - 1) \rightarrow y = 4x + 1$

23. $y - 12 = -\frac{3}{4}(x + 8) \rightarrow y = -\frac{3}{4}x + 6$

24. $m = \frac{-2-3}{6+4} = \frac{-5}{10} = -\frac{1}{2}$

$y - 3 = -\frac{1}{2}(x + 4) \rightarrow y = -\frac{1}{2}x + 1$

25. $m = \frac{4+6}{-3-2} = \frac{10}{-5} = -2$

$y + 6 = -2(x - 2) \rightarrow y = -2x - 2$

26. $(x - 5)(x + 2) =$
$x^2 + 2x - 5x - 10 =$
$x^2 - 3x - 10$

27. $(x - 9)(x - 3) =$
$x^2 - 3x - 9x + 27 =$
$x^2 - 12x + 27$

28. $(x - 4)(x + 8) =$
$x^2 + 8x - 4x - 32 =$
$x^2 + 4x - 32$

29. $(3x - 4)(2x + 5) =$
$6x^2 + 15x - 8x - 20 =$
$6x^2 + 7x - 20$

30. $(5x - 1)(2x - 7) =$
$10x^2 - 35x - 2x + 7 =$
$10x^2 - 37x + 7$

31. $x^2 - 9x + 20 = (x - 5)(x - 4)$

32. $x^2 + 13x + 40 = (x + 5)(x + 8)$

33. $x^2 - x - 12 = (x - 4)(x + 3)$

34. $5x^2 + 8x - 21 = (5x - 7)(x + 3)$

35. $8x^2 + 14x - 15 = (4x - 3)(2x + 5)$

Chapter 2

36. The relation is not a function because $x = 2$ is paired with both $y = 5$ and $y = 7$.

37. The relation is a function because $x^2 - 5y = 20$ is equivalent to $y = \frac{x^2-20}{5}$, which produces one y for each x.

38. $f(-2) =$
$(-2)^3 + 2(-2)^2 - 3(-2) + 1 =$
$-8 + 8 + 6 + 1 = 7$

39. $g(-2) = \sqrt{2(-2) + 3(-2)^2} = \sqrt{-4 + 12} = \sqrt{8} = 2\sqrt{2}$

40. $f(0) = \frac{3(0)-5}{(0)-2} = \frac{5}{2}$

41. $f + g(x) = x + 1 + x^2 - 1 = x^2 + x$,
Domain: all real numbers

42. $g - f(x) = x^2 - 1 - (x + 1) = x^2 - x - 2$,
Domain: all real numbers

43. $f \cdot g(x) = (x + 1)(x^2 - 1) = x^3 + x^2 - x - 1$,
Domain: all real numbers

44. $\frac{f}{g}(x) = \frac{x+1}{x^2-1} = \frac{x+1}{(x+1)(x-1)} = \frac{1}{x-1}$,
Domain: $x \neq \pm 1$

45. $\frac{g}{f}(x) = \frac{x^2-1}{x+1} = \frac{(x+1)(x-1)}{x+1} = x - 1$,
Domain: $x \neq -1$

46. $f \circ g(x) = f(x^2 + 4) = \sqrt{4 + x^2 + 4} = \sqrt{x^2 + 8}$,
Domain: all real numbers

47. $g \circ f(x) = g(\sqrt{4 + x}) = \sqrt{4 + x}^2 + 4 = 4 + x + 4 = x + 8$,
Domain: $x \geq -4$

48. $f \circ h(x) = f(x - 4) = \sqrt{4 + x - 4} = \sqrt{x}$,
Domain: $x \geq 0$

49. $h \circ h(x) = h(x - 4) = x - 4 - 4 = x - 8$,
Domain: all real numbers

50. $f\left(h\left(g\left(x\right)\right)\right)=f\left(h\left(x^2+4\right)\right)=f\left(x^2+4-4\right)=f\left(x^2\right)=\sqrt{4+x^2}$,
    Domain: all real numbers

51. $x=8-3y\rightarrow x-8=-3y\rightarrow y=\frac{x-8}{-3}=\frac{8-x}{3}$,
    Domain: all real numbers

52. $x=\sqrt{y-4}-1\rightarrow x+1=\sqrt{y-4}\rightarrow\left(x+1\right)^2=y-4\rightarrow y=\left(x+1\right)^2+4=x^2+2x+5$,
    Domain: $x\geq-1$

53. $x=\left(y-3\right)^2\rightarrow\sqrt{x}=y-3\rightarrow y=\sqrt{x}+3$,
    Domain: $x\geq0$

54. $x=\frac{2}{y-4}\rightarrow xy-4x=2\rightarrow xy=4x+2\rightarrow y=\frac{4x+2}{x}$,
    Domain: $x\neq0$

55. $x=\frac{y+1}{y^2-1}\rightarrow x=\frac{1}{y-1}\rightarrow xy-x=1\rightarrow xy=x+1\rightarrow y=\frac{x+1}{x}$ or

    $x=\frac{y+1}{y^2-1}\rightarrow x\left(y^2-1\right)=y+1\rightarrow x\left(y+1\right)\left(y-1\right)=y+1\rightarrow y-1=\frac{y+1}{x(y+1)}\rightarrow y-1=\frac{1}{x}\rightarrow y=\frac{1}{x}+1$,
    Domain: $x\neq0$

56. $g(x)=|x-4|+3$

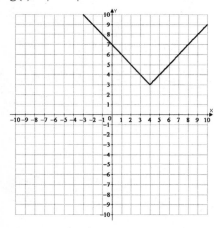

57. $g(x)=2-|x+1|$

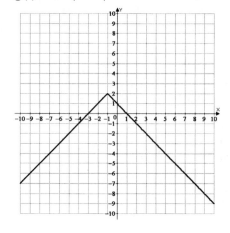

58. $g(x) = 3|4 - x| + 2$

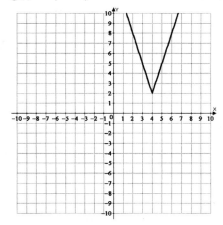

Chapter 3

61. $y < -3x + 7$

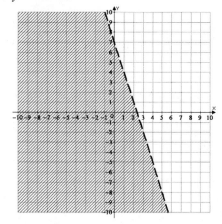

59. $g(x) = -\frac{2}{3}|x| + 7$

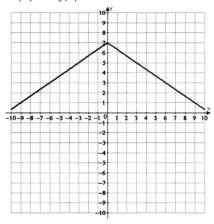

62. $y \geq \frac{5}{3}x - 2$

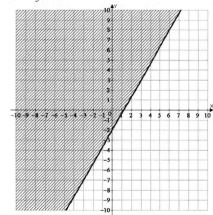

60. $g(x) = 5 - 3|x + 2|$

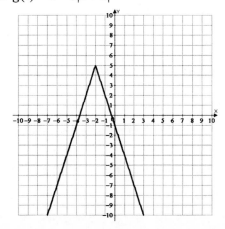

63. $2x + 5y \leq 20$

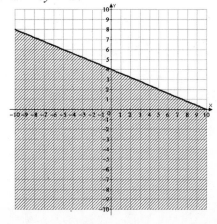

64. $3x - 2y > 12$

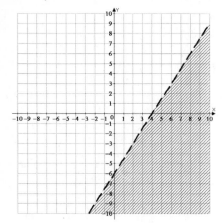

67. $y = -1.39x + 15.30$

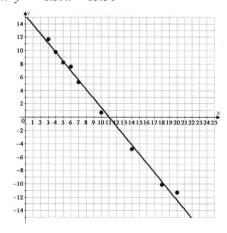

65. $-6y > 12 - 5x$

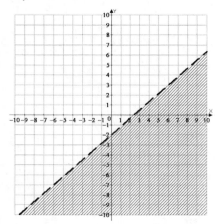

68. $y = 2.68x - 4.22$

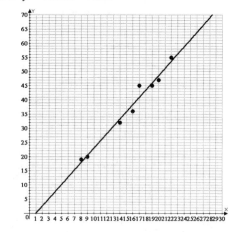

66. $y = -2.61x + 3.16$

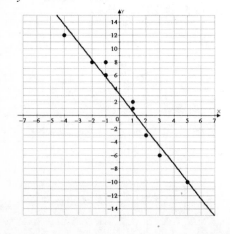

69. $y = -0.46x + 98.09$

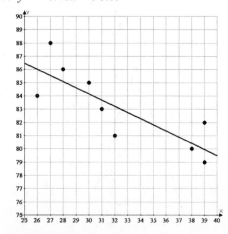

70. $y = 0.08x + 16.98$

73. $y = \begin{cases} x-5 & \text{if } x \geq 0 \\ 2x+1 & \text{if } x < 0 \end{cases}$

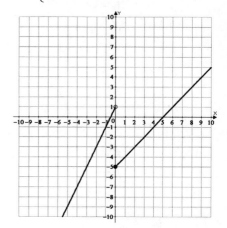

71. $f(x) = \begin{cases} -4 & \text{if } x > 2 \\ 3 & \text{if } x \leq 2 \end{cases}$

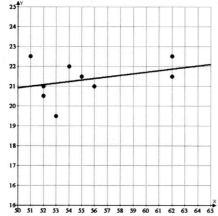

74. $f(x) = \begin{cases} 1-x & \text{if } x < -2 \\ x-1 & \text{if } -2 \leq x \leq 4 \\ 4 & \text{if } x > 4 \end{cases}$

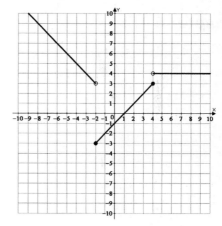

72. $g(x) = \begin{cases} -\frac{2}{3}x+4 & \text{if } x < 2 \\ \frac{1}{5}x-2 & \text{if } x \geq 2 \end{cases}$

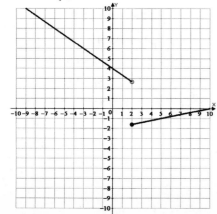

75. $g(x) = \begin{cases} -2x & \text{if } x \le -3 \\ x & \text{if } -3 < x \le -1 \\ 2 & \text{if } -1 < x \le 1 \\ -x & \text{if } 1 < x \le 3 \\ 2x & \text{if } x > 3 \end{cases}$

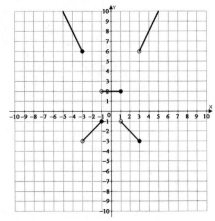

Chapter 4

76. $$|8x + 5| = 19$$

$8x + 5 = 19 \qquad 8x + 5 = -19$

$8x = 14 \qquad\qquad 8x = -24$

$x = \frac{14}{8} = \frac{7}{4} \qquad\quad x = -3$

77. $$|3 - 7x| + 9 = 13$$

$$|3 - 7x| = 4$$

$3 - 7x = 4 \qquad 3 - 7x = -4$

$-7x = 1 \qquad\quad -7x = -7$

$x = \frac{-1}{7} \qquad\qquad x = 1$

78. $$|2x - 11| - 2 = 1$$

$$|2x - 11| = 3$$

$2x - 11 = 3 \qquad 2x - 11 = -3$

$2x = 14 \qquad\quad 2x = 8$

$x = 7 \qquad\qquad x = 4$

79. $$12 - \tfrac{1}{2}|x - 6| = -18$$

$$-\tfrac{1}{2}|x - 6| = -30$$

$$|x - 6| = 60$$

$x - 6 = 60 \qquad x - 6 = -60$

$x = 66 \qquad\qquad x = -54$

80. $$\tfrac{2}{3}|4x + 5| - 18 = 20$$

$$\tfrac{2}{3}|4x + 5| = 38$$

$$|4x + 5| = 57$$

$4x + 5 = 57 \qquad 4x + 5 = -57$

$4x = 52 \qquad\qquad 4x = -62$

$x = 13 \qquad\qquad x = -15.5$

81. $$|4 - x| \le 12$$

$$-12 \le 4 - x \le 12$$

$$-16 \le -x \le 8$$

$$16 \ge x \ge -8$$

82. $$|3x - 11| + 2 > 36$$

$$|3x - 11| > 34$$

$-34 > 3x - 11 \qquad\quad 3x - 11 > 34$

$-23 > 3x \qquad or \qquad 3x > 45$

$\tfrac{-23}{3} > x \qquad\qquad\qquad x > 15$

83. $$3 - 8|5t - 11| < -29$$

$$-8|5t - 11| < -32$$

$$|5t - 11| > 4$$

$-4 > 5t - 11 \qquad\quad 5t - 11 > 4$

$7 > 5t \qquad or \qquad 5t > 15$

$\tfrac{7}{5} > t \qquad\qquad\qquad t > 3$

84.
$$14 - \left| \tfrac{1}{2}x + 7 \right| \geq 2$$
$$-\left| \tfrac{1}{2}x + 7 \right| \geq -12$$
$$\left| \tfrac{1}{2}x + 7 \right| \leq 12$$
$$-12 \leq \tfrac{1}{2}x + 7 \leq 12$$
$$-19 \leq \tfrac{1}{2}x \leq 5$$
$$-38 \leq x \leq 10$$

85.
$$3\left| 2 - 9x \right| + 2 < 50$$
$$3\left| 2 - 9x \right| < 48$$
$$\left| 2 - 9x \right| < 16$$
$$-16 < 2 - 9x < 16$$
$$-18 < -9x < 14$$
$$2 > x > -\tfrac{14}{9}$$

86. $f(x) = \tfrac{1}{2}|x + 5|$

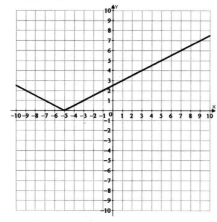

87. $g(x) = |x - 4| - 7$

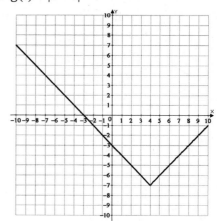

88. $y = 1 - |x + 4|$

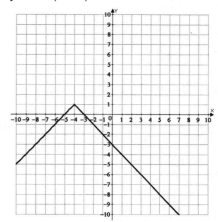

89. $f(x) = 2|x - 3| + 5$

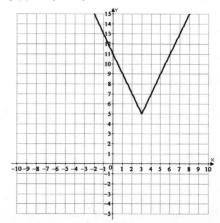

90. $g(x) = 4 - \frac{1}{2}|x+6|$

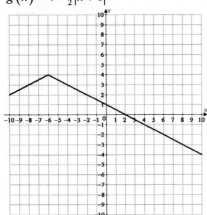

Chapter 5

91. $x = 3, y = -1$

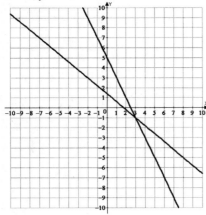

92. $x = 4, y = -4$

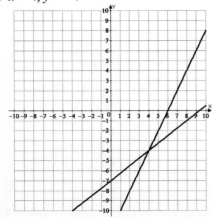

93. $x = -2, y = -4$

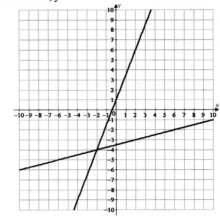

94. $x = 6, y = -2$

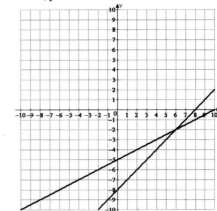

95. $x = -5, y = 2$

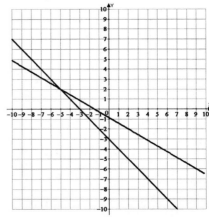

96. $y = \frac{1}{2}x + 6$

$2x - 3\left(\frac{1}{2}x + 6\right) = -20$

$2x - 1.5x - 18 = -20$

$0.5x = -2$

$x = -4$

$y = 4$

97. $\qquad x = 11 - y$

$2x - 4y = -2$

$2(11 - y) - 4y = -2$

$22 - 2y - 4y = -2$

$22 - 6y = -2$

$-6y = -24$

$y = 4$

$x = 7$

98. $\qquad y = 3x + 9$

$4x + 5(3x + 9) = 7$

$4x + 15x + 45 = 7$

$19x = -38$

$x = -2$

$y = 3$

99. $\qquad y = 2 - 5x$

$3x - 2(2 - 5x) = -17$

$3x - 4 + 10x = -17$

$13x = -13$

$x = -1$

$y = 7$

100. $\qquad x = 7y - 31$

$5(7y - 31) - y = 15$

$35y - 155 - y = 15$

$34y = 170$

$y = 5$

$x = 4$

101. $2(4x + 3y) = 2(14)$

$3(9x - 2y) = 3(14)$

$8x + 6y = 28$

$\underline{27x - 6y = 42}$

$35x = 70$

$x = 2$

$4(2) + 3y = 14$

$8 + 3y = 14$

$3y = 6$

$y = 2$

102. $-2(2x - 5y) = -2(16)$

$-4x + 10y = -32$

$\underline{4x - 3y = 11}$

$7y = -21$

$y = -3$

$4x - 3(-3) = 11$

$4x + 9 = 11$

$4x = 2$

$x = \frac{1}{2}$

103. $2(5x - 3y) = 2(-11)$

$10x - 6y = -22$

$\underline{7x + 6y = -12}$

$17x = -34$

$x = -2$

$7(-2) + 6y = -12$

$-14 + 6y = -12$

$6y = 2$

$y = \frac{1}{3}$

104. $-5(7x - 6y) = -5(13)$

$6(6x - 5y) = 6(11)$

$-35x + 30y = -65$

$\underline{36x - 30y = 66}$

$x = 1$

$6(1) - 5y = 11$

$-5y = 5$

$y = -1$

105. Dependent system—infinitely many solutions.

$2(3x - 2y) = 2(1)$

$6x - 4y = 2$

$\underline{-6x + 4y = -2}$

$0 = 0$

106. $2(x + y - z) = 2(6)$

$2x + 2y - 2z = 12$

$\underline{x - y + 2z = -7}$

$3x + y = 5$

$x + y - z = 6$

$x + 2y + z = 2$

$2x + 3y = 8$

$-3(3x + y) = -3(5)$

$-9x - 3y = -15$

$\underline{2x + 3y = 8}$

$-7x = -7$

$x = 1$

$2(1) + 3y = 8$

$y = 2$

$z = 1$

107. $x + y + z = 6$

$\underline{2x - y + z = 3}$

$3x + 2z = 9$

$2(2x - y + z) = 2(3)$

$4x - 2y + 2z = 6$

$\underline{x + 2y - 3z = -4}$

$5x - z = 2$

$z = 5x - 2$

$3x + 2(5x - 2) = 9$

$3x + 10x - 4 = 9$

$13x = 13$

$x = 1$

$z = 3$

$y = 2$

108. $6x - 2y + 4z = 2$

$2(4x + y + 2z) = 2(8)$

$6x - 2y + 4z = 2$

$\underline{8x + 2y + 4z = 16}$

$14x + 8z = 18$

$5x - 3y + 6z = 5$

$3(4x + y + 2z) = 3(8)$

$5x - 3y + 6z = 5$

$\underline{12x + 3y + 6z = 24}$

$17x + 12z = 29$

$3(14x + 8z) = 3(18)$

$\underline{-2(17x + 12z) = -2(29)}$

$42x + 24z = 54$

$\underline{-34x - 24z = -58}$

$8x = -4$

$x = -\frac{1}{2}$

$z = \frac{25}{8}$

$y = \frac{15}{4}$

109. $3x - 2y + z = 2$

$\underline{x + 4y - z = 4}$

$4x + 2y = 6$

$2x - 3y + z = 1$

$\underline{x + 4y - z = 4}$

$3x + y = 5$

$4x + 2y = 6$

$-2(3x + y) = -2(5)$

$4x + 2y = 6$

$\underline{-6x - 2y = -10}$

$-2x = -4$

$x = 2$

$y = -1$

$z = -6$

110. No solution.

$$3x + 5y - 2z = 10 \qquad 6x + 10y - 4z = 2$$
$$2(x - 8y + z) = 2(4) \qquad 4(x - 8y + z) = 4(4)$$
$$3x + 5y - 2z = 10 \qquad 6x + 10y - 4z = 2$$
$$\underline{2x - 16y + 2z = 8} \qquad \underline{4x - 32y + 4z = 16}$$
$$5x - 11y = 18 \qquad 10x - 22y = 18$$
$$2(5x - 11y) = 2(18)$$
$$10x - 22y = 36$$
$$\underline{10x - 22y = 18}$$
$$0 = 54$$

111. $y \le 6 - x$
     $2x - 3y > 2$

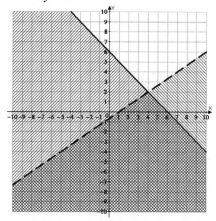

112. $y < \frac{1}{2}x + 1$
     $x - 3 \le y$

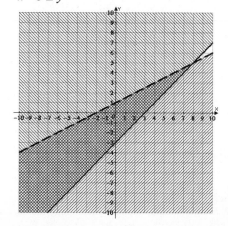

113. $x - 2y > 12$
     $3x + y < 15$

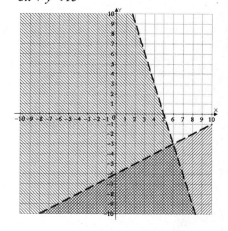

114. $y \le 5x + 17$
     $7x + 33 \ge 6y$

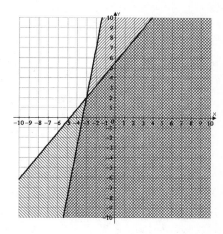

115. $y > \frac{3}{4}x - 3$

$2x - 5y \le 1$

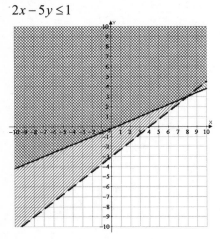

116. Maximum profit = $110

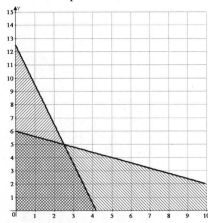

x	y	Profit
0	0	0
0	6	12(0) + 16(6) = 96
2.5	5	12(2.5) + 16(5) = 110
4.16	0	12(4.16) + 16(0) = 50

117. Maximum profit = $100

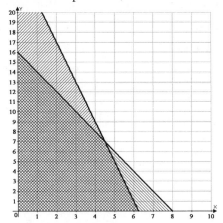

x	y	Profit
0	0	0
0	16	11.50(0) + 6.25(16) = 100
4.5	7	11.50(4.5) + 6.25(7) = 95.50
6.25	0	11.50(6.25) + 6.25(0) = 71.875

118. Minimum cost = $36

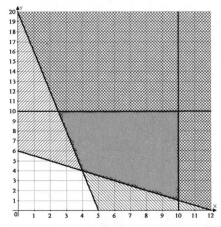

x	y	Cost
4	4	4(4) + 5(4) = 36
2.5	10	4(2.5) + 5(10) = 60
10	10	4(10) + 5(10) = 90
10	1	4(10) + 5(1) = 45

119. Minimum cost = $350

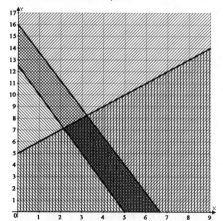

x	y	Cost
5	0	70(5) + 60(0) = 350
2.1	7.1	70(2.1) + 60(0) = 573
3.2	8.2	70(3.2) + 60(8.2) = 716
6.6	0	70(6.6) + 60(0) = 462

120. Maximum profit = $23,333.33

$$x + y \leq 100$$
$$5x + 2y \leq 400$$
$$2x + 5y \leq 400$$
$$30x + 10y \geq 500$$
$$\text{Profit} = 200x + 250y$$

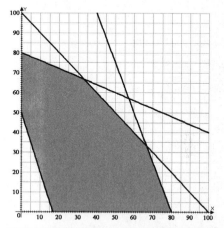

x	y	Profit
0	50	200(0) + 250(50) = 12,500
0	80	200(0) + 250(80) = 20,000
33.3	66.6	200(33.3) + 250(66.6) = 23,333.33
66.6	33.3	200(66.6) + 250(33.3) = 21,666.66
80	0	200(80) + 250(0) = 16,000
50	0	200(50) + 250(0) = 10,000

Chapter 6

121. $[F] - [A] = \begin{bmatrix} -1 & 0 \\ -1 & -1 \end{bmatrix}$

122. $[B] \cdot [D] = \begin{bmatrix} -2 & -8 & -5 \\ -3 & -13 & -8 \\ 0 & 2 & 1 \end{bmatrix}$

123. $[A] \cdot [D] = \begin{bmatrix} -3 & -7 & -5 \\ 2 & 6 & 4 \end{bmatrix}$

124. $[F] + [D]$ is not possible.

125. $[E] \cdot [B] \cdot [C] = \begin{bmatrix} 0 & 0 \end{bmatrix} \times \begin{bmatrix} 2 \\ -1 \end{bmatrix} = [0]$

126. $\begin{vmatrix} 2 & 1 \\ 3 & 4 \end{vmatrix} = 8 - 3 = 5$

127. $\begin{vmatrix} -3 & 2 \\ 6 & -4 \end{vmatrix} = 12 - 12 = 0$

128. $\begin{vmatrix} 8 & -3 \\ -2 & 5 \end{vmatrix} = 40 - 6 = 34$

129. $\begin{vmatrix} 2 & -1 & 0 \\ 1 & 0 & -2 \\ 0 & 1 & 2 \end{vmatrix} = 6$

130. $\begin{vmatrix} 1 & 3 & 7 \\ -2 & 6 & 4 \\ 3 & 7 & -1 \end{vmatrix} = -228$

131. $\begin{vmatrix} 2 & -3 \\ 4 & -2 \end{vmatrix} = 8,\ \begin{vmatrix} 3 & -3 \\ 10 & -2 \end{vmatrix} = 24,\ \begin{vmatrix} 2 & 3 \\ 4 & 10 \end{vmatrix} = 8,$
$x = \frac{24}{8} = 3, y = \frac{8}{8} = 1$

132. $\begin{vmatrix} 3 & 1 \\ -3 & 2 \end{vmatrix} = 9,\ \begin{vmatrix} -2 & 1 \\ -4 & 2 \end{vmatrix} = 0,\ \begin{vmatrix} 3 & -2 \\ -3 & -4 \end{vmatrix} = -18,$
$x = \frac{0}{9} = 0, y = \frac{-18}{9} = -2$

133. $\begin{vmatrix} 4 & -7 \\ 5 & 2 \end{vmatrix} = 43,\ \begin{vmatrix} 3 & -7 \\ -3 & 2 \end{vmatrix} = -15,\ \begin{vmatrix} 4 & 3 \\ 5 & -3 \end{vmatrix} = -27,$
$x = \frac{-15}{43}, y = \frac{-27}{43}$

134. $\begin{vmatrix} 2 & -3 & 1 \\ 3 & -1 & -1 \\ 4 & -6 & 2 \end{vmatrix} = 0$, no solution

135. $\begin{vmatrix} 2 & -1 & 3 \\ 1 & -5 & -2 \\ -4 & -2 & 1 \end{vmatrix} = -91,\ \begin{vmatrix} 4 & -1 & 3 \\ 1 & -5 & -2 \\ 3 & -2 & 1 \end{vmatrix} = 10,$
$\begin{vmatrix} 2 & 4 & 3 \\ 1 & 1 & -2 \\ -4 & 3 & 1 \end{vmatrix} = 63,\ \begin{vmatrix} 2 & -1 & 4 \\ 1 & -5 & 1 \\ -4 & -2 & 3 \end{vmatrix} = -107,$
$x = -\frac{10}{91}, y = -\frac{63}{91}, z = \frac{107}{91}$

136. No solution.
$\begin{bmatrix} 5 & -2 & | & 4 \\ -10 & 4 & | & 1 \end{bmatrix} \rightarrow \begin{bmatrix} 1 & -0.4 & | & 0.84 \\ 0 & 0 & | & 9.4 \end{bmatrix}$

137. $\begin{bmatrix} 3 & -4 & | & 7 \\ 6 & -2 & | & 5 \end{bmatrix} \rightarrow \begin{bmatrix} 1 & 0 & | & \frac{1}{3} \\ 0 & 1 & | & -\frac{3}{2} \end{bmatrix}$

138. $\begin{bmatrix} 2 & -3 & | & 4 \\ 4 & -5 & | & 3 \end{bmatrix} \rightarrow \begin{bmatrix} 1 & 0 & | & -5.5 \\ 0 & 1 & | & -5 \end{bmatrix}$

139. $\begin{bmatrix} 1 & 2 & 1 & | & 2 \\ 1 & 1 & -1 & | & 6 \\ 1 & -1 & 1 & | & -4 \end{bmatrix} \rightarrow \begin{bmatrix} 1 & 0 & 0 & | & 1 \\ 0 & 1 & 0 & | & 2 \\ 0 & 0 & 1 & | & -3 \end{bmatrix}$

140. $\begin{bmatrix} 1 & 1 & 0 & | & 6 \\ 2 & 0 & 1 & | & 3 \\ 0 & 2 & -3 & | & -4 \end{bmatrix} \rightarrow \begin{bmatrix} 1 & 0 & 0 & | & -1.75 \\ 0 & 1 & 0 & | & 7.75 \\ 0 & 0 & 1 & | & 6.5 \end{bmatrix}$

141. $\begin{bmatrix} 2 & 1 \\ 4 & 3 \end{bmatrix}^{-1} = \frac{1}{2}\begin{bmatrix} 3 & -1 \\ -4 & 2 \end{bmatrix} = \begin{bmatrix} \frac{3}{2} & -\frac{1}{2} \\ -2 & 1 \end{bmatrix}$

142. $\begin{bmatrix} 6 & 1 \\ 2 & \frac{1}{2} \end{bmatrix}^{-1} = \begin{bmatrix} \frac{1}{2} & -1 \\ -2 & 6 \end{bmatrix}$

143. $\begin{bmatrix} 4 & 4 \\ 1 & 1 \end{bmatrix}$ No inverse; determinant = 0.

144. $\begin{bmatrix} 2 & 0 & -1 \\ 3 & 1 & 2 \\ 5 & -2 & 1 \end{bmatrix}^{-1} = \begin{bmatrix} \frac{5}{21} & \frac{2}{21} & \frac{1}{21} \\ \frac{1}{3} & \frac{1}{3} & -\frac{1}{3} \\ -\frac{11}{21} & \frac{4}{21} & \frac{2}{21} \end{bmatrix}$

145. $\begin{bmatrix} 1 & 3 & 7 \\ -2 & 6 & 4 \\ 3 & 7 & -1 \end{bmatrix}^{-1} = \begin{bmatrix} \frac{11}{102} & \frac{-23}{102} & \frac{5}{34} \\ \frac{-7}{102} & \frac{5}{51} & \frac{3}{34} \\ \frac{8}{51} & \frac{-1}{102} & \frac{-1}{17} \end{bmatrix}$

146. $\begin{bmatrix} x \\ y \end{bmatrix} = \begin{bmatrix} 3 & 1 \\ 2 & 1 \end{bmatrix}^{-1} \cdot \begin{bmatrix} 2 \\ 0 \end{bmatrix} = \begin{bmatrix} 2 \\ -4 \end{bmatrix}$

147. $\begin{bmatrix} x \\ y \end{bmatrix} = \begin{bmatrix} 3 & -7 \\ -4 & 6 \end{bmatrix}^{-1} \cdot \begin{bmatrix} 2 \\ -6 \end{bmatrix} = \begin{bmatrix} 3 \\ 1 \end{bmatrix}$

148. $\begin{bmatrix} x \\ y \end{bmatrix} = \begin{bmatrix} 5 & -2 \\ 3 & 1 \end{bmatrix}^{-1} \cdot \begin{bmatrix} 7 \\ 2 \end{bmatrix} = \begin{bmatrix} 1 \\ -1 \end{bmatrix}$

149. $\begin{bmatrix} x \\ y \\ z \end{bmatrix} = \begin{bmatrix} 5 & -3 & -6 \\ 4 & -6 & -3 \\ -1 & 9 & 9 \end{bmatrix}^{-1} \cdot \begin{bmatrix} 5 \\ 4 \\ 7 \end{bmatrix} = \begin{bmatrix} 2 \\ \frac{1}{3} \\ \frac{2}{3} \end{bmatrix}$

150. $\begin{bmatrix} x \\ y \\ z \end{bmatrix} = \begin{bmatrix} 4 & -6 & 8 \\ 5 & 1 & -2 \\ 6 & -8 & 12 \end{bmatrix}^{-1} \cdot \begin{bmatrix} 4 \\ 4 \\ 6 \end{bmatrix} = \begin{bmatrix} \frac{5}{6} \\ 0 \\ \frac{1}{12} \end{bmatrix}$

Chapter 7

151. $27^{\frac{2}{3}} = 9$

152. $(16)^{\frac{3}{4}} = 8$

153. $\left(\frac{81}{49}\right)^{-\frac{1}{2}} = \left(\frac{49}{81}\right)^{\frac{1}{2}} = \frac{7}{9}$

154. $\left(\frac{8}{1,000}\right)^{-\frac{5}{3}} = \left(\frac{1,000}{8}\right)^{\frac{5}{3}} = \left(\frac{10}{2}\right)^5 = 5^5 = 3,125$

155. $(-32)^{\frac{4}{5}} = (-2)^4 = 16$

156. $\sqrt{81a^4 b^8 c^7} = 9a^2 b^4 c^3 \sqrt{c}$

157. $\sqrt[3]{81x^3 y^9 z^4} = 3xy^3 z \sqrt[3]{3z}$

158. $\left(\frac{16a^5 b^4}{27z^9}\right)^{\frac{1}{3}} = \frac{2ab\sqrt[3]{2a^2 b}}{3z^3}$

159. $\sqrt[5]{\frac{128x^3 y^7}{z^{12}}} = \frac{2y}{z^2} \sqrt[5]{\frac{2x^3 y^2}{z^2}}$

160. $\left(\frac{9x^7 y^3}{125z^4}\right)^{-\frac{3}{2}} = \left(\frac{125z^4}{9x^7 y^3}\right)^{\frac{3}{2}} = \frac{125^{\frac{3}{2}} z^6}{9^{\frac{3}{2}} x^{\frac{21}{2}} y^{\frac{9}{2}}} = \frac{5z^3 \sqrt{5}}{27x^{10} y^4 \sqrt{xy}} = \frac{5z^3 \sqrt{5}}{27x^{10} y^4 \sqrt{xy}} \cdot \frac{\sqrt{xy}}{\sqrt{xy}} = \frac{5z^3 \sqrt{5xy}}{27x^{10} y^4 \cdot xy} = \frac{5z^3 \sqrt{5xy}}{27x^{11} y^5}$

161. $\frac{\sqrt{x}}{\sqrt{x}+3} \cdot \frac{\sqrt{x}-3}{\sqrt{x}-3} = \frac{x-3\sqrt{x}}{x-9}$

162. $\frac{\sqrt{2}+7}{\sqrt{2}-7} \cdot \frac{\sqrt{2}+7}{\sqrt{2}+7} = \frac{2+14\sqrt{2}+49}{2-49} = \frac{51+14\sqrt{2}}{-47} = -\frac{51+14\sqrt{2}}{47}$

163. $\frac{2\sqrt{3}-\sqrt{7}}{3\sqrt{3}+\sqrt{7}} \cdot \frac{3\sqrt{3}-\sqrt{7}}{3\sqrt{3}-\sqrt{7}} = \frac{18-3\sqrt{21}-2\sqrt{21}+7}{27-7} = \frac{25-5\sqrt{21}}{20} = \frac{5(5-\sqrt{21})}{20} = \frac{5-\sqrt{21}}{4}$

164. $\dfrac{3}{\sqrt[3]{4}} \cdot \dfrac{\left(\sqrt[3]{4}\right)^2}{\left(\sqrt[3]{4}\right)^2} = \dfrac{\sqrt[3]{16}}{4} = \dfrac{2\sqrt[3]{2}}{4} = \dfrac{\sqrt[3]{2}}{2}$

165. $\dfrac{1+\sqrt[3]{4}}{\sqrt[3]{4}} \cdot \dfrac{\left(\sqrt[3]{4}\right)^2}{\left(\sqrt[3]{4}\right)^2} = \dfrac{\left(\sqrt[3]{4}\right)^2+4}{4} = \dfrac{\sqrt[3]{16}+4}{4} = \dfrac{2\sqrt[3]{2}+4}{4} = \dfrac{2\left(\sqrt[3]{2}+2\right)}{4} = \dfrac{\sqrt[3]{2}+2}{2}$

166. $\sqrt{4x+5}+2=7$

$\qquad \sqrt{4x+5}=5$

$\qquad 4x+5=25$

$\qquad 4x=20$

$\qquad x=5$

167. $\qquad t-6=\sqrt{t-4}$

$\qquad (t-6)^2 = t-4$

$\qquad t^2-12t+36=t-4$

$\qquad t^2-13t+40=0$

$\qquad (t-8)(t-5)=0$

$\qquad t=8 \qquad t=5$

168. $\sqrt{x+2}^{\,2} = \left(\sqrt{x+3}-1\right)^2$

$\qquad x+2=x+3-2\sqrt{x+3}+1$

$\qquad -2=-2\sqrt{x+3}$

$\qquad 1=\sqrt{x+3}$

$\qquad 1=x+3$

$\qquad x=-2$

169. $\sqrt[3]{4x+5}=3$

$\qquad 4x+5=27$

$\qquad 4x=22$

$\qquad x=5.5$

170. $\sqrt[4]{6x+7}-\sqrt[4]{x+2}=0$

$\qquad \sqrt[4]{6x+7}=\sqrt[4]{x+2}$

$\qquad 6x+7=x+2$

$\qquad 5x=-5$

$\qquad x=-1$

171. $f(x)=\sqrt{x+2}-4$

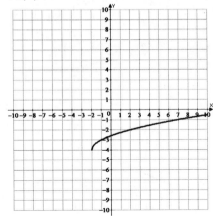

172. $y=-2\sqrt{x-5}$

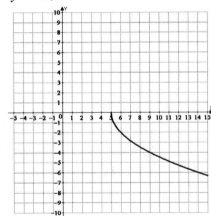

173. $y=9-\sqrt{3-x}$

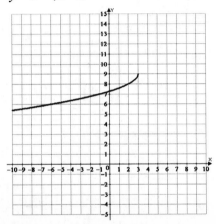

174. $f(x) = 4\sqrt[3]{x} + 1$

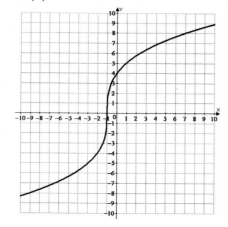

175. $g(x) = 3 - \frac{1}{2}\sqrt[3]{x-5}$

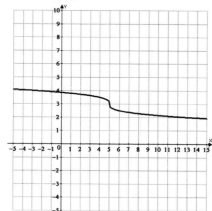

Chapter 8

176. $4x^2 - 64 = 0$

$x^2 - 16 = 0$

$x^2 = 16$

$x = \pm 4$

177. $(2x-3)^2 = 25$

$2x - 3 = \pm 5$

$2x - 3 = 5 \qquad 2x - 3 = -5$

$2x = 8 \qquad 2x = -2$

$x = 4 \qquad x = -1$

178. $(3x+2)^2 = 48$

$3x + 2 = \pm\sqrt{48} = \pm 4\sqrt{3}$

$3x = -2 \pm 4\sqrt{3}$

$x = \frac{-2 \pm 4\sqrt{3}}{3}$

179. $x^2 - 6x + 9 = 25$

$(x-3)^2 = 25$

$x - 3 = \pm 5$

$x - 3 = 5 \qquad x - 3 = -5$

$x = 8 \qquad x = -2$

180. $(x-7)^2 + 5 = 86$

$(x-7)^2 = 81$

$x - 7 = \pm 9$

$x - 7 = 9 \qquad x - 7 = -9$

$x = 16 \qquad x = -2$

181. $x^2 + 6x = 3$

$x^2 + 6x + 9 = 3 + 9$

$(x+3)^2 = 12$

$x + 3 = \pm\sqrt{12} = \pm 2\sqrt{3}$

$x = -3 \pm 2\sqrt{3}$

182. $x^2 + 3x - 4 = 0$

$x^2 + 3x + \frac{9}{4} = 4 + \frac{9}{4}$

$\left(x + \frac{3}{2}\right)^2 = \frac{25}{4}$

$x + \frac{3}{2} = \pm\frac{5}{2}$

$x + \frac{3}{2} = \frac{5}{2} \qquad x + \frac{3}{2} = -\frac{5}{2}$

$x = \frac{2}{2} = 1 \qquad x = -\frac{8}{2} = -4$

183. $$5x^2 - 3x - 2 = 0$$
$$x^2 - \tfrac{3}{5}x - \tfrac{2}{5} = 0$$
$$x^2 - \tfrac{3}{5}x = \tfrac{2}{5}$$
$$x^2 - \tfrac{3}{5}x + \tfrac{9}{100} = \tfrac{2}{5} + \tfrac{9}{100}$$
$$\left(x - \tfrac{3}{10}\right)^2 = \tfrac{49}{100}$$
$$x - \tfrac{3}{10} = \pm\tfrac{7}{10}$$
$$x - \tfrac{3}{10} = \tfrac{7}{10} \qquad x - \tfrac{3}{10} = -\tfrac{7}{10}$$
$$x = 1 \qquad x = -\tfrac{4}{10} = -\tfrac{2}{5}$$

185. $$3x^2 - 10x + 7 = 0$$
$$3x^2 - 10x = -7$$
$$x^2 - \tfrac{10}{3}x = -\tfrac{7}{3}$$
$$x^2 - \tfrac{10}{3}x + \tfrac{25}{9} = -\tfrac{7}{3} + \tfrac{25}{9}$$
$$\left(x - \tfrac{5}{3}\right)^2 = \tfrac{4}{9}$$
$$x - \tfrac{5}{3} = \pm\tfrac{2}{3}$$
$$x - \tfrac{5}{3} = \tfrac{2}{3} \qquad x - \tfrac{5}{3} = -\tfrac{2}{3}$$
$$x = \tfrac{7}{3} \qquad x = 1$$

184. $$x^2 + 5x - 2 = 0$$
$$x^2 + 5x + \tfrac{25}{4} = 2 + \tfrac{25}{4}$$
$$\left(x + \tfrac{5}{2}\right)^2 = \tfrac{33}{4}$$
$$x + \tfrac{5}{2} = \pm\tfrac{\sqrt{33}}{2}$$
$$x = \tfrac{-5 \pm \sqrt{33}}{2}$$

186. $$x = \tfrac{-7 \pm \sqrt{7^2 - 4\cdot6\cdot2}}{2\cdot6} = \tfrac{-7 \pm \sqrt{49-48}}{12} = \tfrac{-7\pm1}{12}$$
$$x = \tfrac{-8}{12} = \tfrac{-2}{3} \qquad x = \tfrac{-6}{12} = -\tfrac{1}{2}$$

187. $$\tfrac{1}{2}x^2 + x = \tfrac{1}{3} \rightarrow \tfrac{1}{2}x^2 + x - \tfrac{1}{3} = 0$$
$$x = \tfrac{-1 \pm \sqrt{1^2 - 4\left(\tfrac{1}{2}\right)\left(\tfrac{-1}{3}\right)}}{2\left(\tfrac{1}{2}\right)} = -1 \pm \sqrt{1 + \tfrac{2}{3}} = -1 \pm \sqrt{\tfrac{5}{3}} = -1 \pm \tfrac{\sqrt{15}}{3} = \tfrac{-3 \pm \sqrt{15}}{3}$$

188. $$z = \tfrac{-7 \pm \sqrt{7^2 - 4\cdot6\cdot(-5)}}{2\cdot6} = \tfrac{-7 \pm \sqrt{49+120}}{12} = \tfrac{-7 \pm \sqrt{169}}{12} = \tfrac{-7\pm13}{12}$$
$$z = \tfrac{6}{12} = \tfrac{1}{2} \qquad z = \tfrac{-20}{12} = \tfrac{-5}{3}$$

189. $$x = \tfrac{-0.06 \pm \sqrt{(0.06)^2 - 4(0.01)(-0.08)}}{2(0.01)} = \tfrac{-0.06 \pm \sqrt{0.0036+0.0032}}{0.02} = \tfrac{-0.06 \pm \sqrt{0.0068}}{0.02} = \tfrac{-0.06 \pm \sqrt{0.0068}}{0.02} \approx \tfrac{-0.06 \pm 0.08246}{0.02}$$
$$x \approx \tfrac{0.02246}{0.02} \approx 1.123 \qquad x \approx \tfrac{-0.14246}{0.02} \approx -7.123$$

190. $$2x - 3 = 3x^2 \rightarrow 3x^2 - 2x + 3 = 0$$
$$x = \tfrac{2 \pm \sqrt{(-2)^2 - 4\cdot2\cdot3}}{2\cdot2} = \tfrac{2 \pm \sqrt{4-24}}{4} = \tfrac{2 \pm i\sqrt{20}}{4} = \tfrac{2 \pm 2i\sqrt{5}}{4} = \tfrac{1 \pm i\sqrt{5}}{2}$$

191. $b^2 - 4ac = 25 - 4(-3) = 37$, two real solutions

192. $b^2 - 4ac = 64 - 4 \cdot 3 = 52$, two real solutions

193. $b^2 - 4ac = 9 - 4 \cdot 4 \cdot 5 = -71$, two nonreal solutions

194. $b^2 - 4ac = 144 - 4 \cdot 5(-1) = 164$, two real solutions

195. $b^2 - 4ac = 25 - 4 \cdot 7 \cdot 2 = 25 - 56 = -31$, two nonreal solutions

196. $(7 + 5i) - (5 - 3i) + (1 + 4i) = 2 + 8i + 1 + 4i = 3 + 12i$

197. $(3 + 2i)(2 - i) = 6 - 3i + 4i - 2i^2 = 8 + i$

203. $g(x) = 3(x - 4)^2 + 1$

198. $\frac{-2i}{5-3i} \cdot \frac{5+3i}{5+3i} = \frac{-10i-6i^2}{25-9i^2} = \frac{6-10i}{25+9} = \frac{6-10i}{34}$

199. $x = \frac{2\pm\sqrt{4-4\cdot1\cdot5}}{2\cdot1} = \frac{2\pm\sqrt{-16}}{2} = \frac{2\pm4i}{2} = 1 \pm 2i$

200. $x = \frac{-3\pm\sqrt{9-4\cdot2\cdot3}}{2\cdot2} = \frac{-3\pm\sqrt{-15}}{4} = \frac{-3\pm i\sqrt{15}}{4}$

201. $f(x) = 2(x + 1)^2$

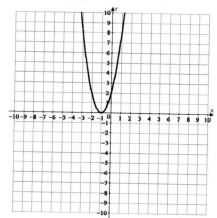

202. $y = -\frac{1}{2}(x - 3)^2 + 2$

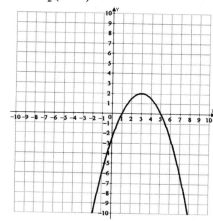

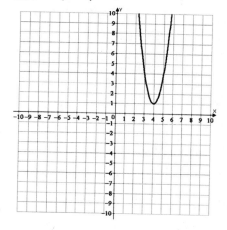

204. $y = x^2 - 4x + 7$

$y - 7 + 4 = x^2 - 4x + 4$

$y - 3 = (x - 2)^2$

$y = (x - 2)^2 + 3$

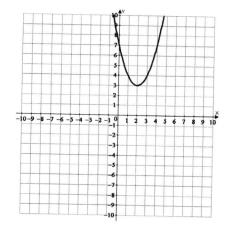

205. $f(x) = -2x^2 + 12x - 10$

$y + 10 = -2x^2 + 12x$

$\frac{y+10}{-2} = \frac{-2x^2+12x}{-2}$

$\frac{y+10}{-2} + 9 = x^2 - 6x + 9$

$\frac{y+10}{-2} + \frac{-18}{-2} = (x-3)^2$

$\frac{y-8}{-2} = (x-3)^2$

$y = -2(x-3)^2 + 8$

206. $y = a(x-2)^2 + 6$

$-3 = a(5-2)^2 + 6$

$-9 = 9a$

$a = -1$

$y = -(x-2)^2 + 6$

207. $y = a(x+5)^2 + 4$

$12 = a(-1+5)^2 + 4$

$8 = 16a$

$a = \frac{1}{2}$

$y = \frac{1}{2}(x+5)^2 + 4$

208. $y = a(x-1)^2 - 3$

$-5 = a(0-1)^2 - 3$

$-2 = a$

$y = -2(x-1)^2 - 3$

209. $y = x^2 - 5x + 3$

$17 = a(-2)^2 + b(-2) + c$ $4a - 2b + c = 17$ $a = 1$

$-1 = a(1)^2 + b(1) + c \rightarrow a + b + c = -1 \rightarrow b = -5$

$9 = a(6)^2 + b(6) + c$ $36a + 6b + c = 9$ $c = 3$

210. $y = -3x^2 + 7x + 8$

$-18 = a(-2)^2 + b(-2) + c$ $4a - 2b + c = -18$ $a = -3$

$-2 = a(-1)^2 + b(-1) + c \rightarrow a - b + c = -2 \rightarrow b = 7$

$10 = a(2)^2 + b(2) + c$ $4a + 2b + c = 10$ $c = 8$

Chapter 9

211. $x^2 + 6x - 3y = 3$

$x^2 + 6x + 9 = 3y + 3 + 9$

$(x+3)^2 = 3y + 12$

$\frac{1}{3}(x+3)^2 = y + 4$

$\frac{1}{3}(x+3)^2 - 4 = y$

212. $5y - x^2 = 4x - 16$

$5y + 16 = x^2 + 4x$

$5y + 16 + 4 = x^2 + 4x + 4$

$5y + 20 = (x+2)^2$

$y + 4 = \frac{1}{5}(x+2)^2$

$y = \frac{1}{5}(x+2)^2 - 4$

213. $5y^2 + 20y + 50 = 10x$

$\quad 5y^2 + 20y = 10x - 50$

$\quad y^2 + 4y = 2x - 10$

$\quad y^2 + 4y + 4 = 2x - 10 + 4$

$\quad (y+2)^2 = 2x - 6$

$\quad \frac{1}{2}(y+2)^2 = x - 3$

$\quad \frac{1}{2}(y+2)^2 + 3 = x$

214. $2x^2 + y - 34x + 99 = 0$

$\quad 2x^2 - 34x = -y - 99$

$\quad x^2 - 17x = \frac{-y-99}{2}$

$\quad x^2 - 17x + \left(\frac{17}{2}\right)^2 = \frac{-y-99}{2} + \left(\frac{17}{2}\right)^2$

$\quad \left(x - \frac{17}{2}\right)^2 = \frac{-2y-198}{4} + \frac{289}{4}$

$\quad \left(x - \frac{17}{2}\right)^2 = \frac{-2y+100}{4}$

$\quad 4\left(x - \frac{17}{2}\right)^2 = -2y + 100$

$\quad -2\left(x - \frac{17}{2}\right)^2 = y - 50$

$\quad -2\left(x - \frac{17}{2}\right)^2 + 50 = y$

215. $2y^2 + 4y - x - 9 = 0$

$\quad 2y^2 + 4y = x + 9$

$\quad y^2 + 2y + 1 = \frac{x+9}{2} + 1$

$\quad (y+1)^2 = \frac{x+11}{2}$

$\quad 2(y+1)^2 = x + 11$

$\quad 2(y+1)^2 - 11 = x$

216. $y = -2(x-4)^2 - 3$

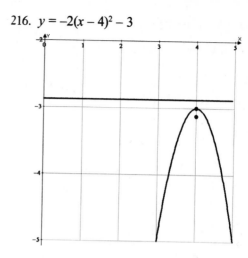

217. $x + 1 = 2(y-3)^2$

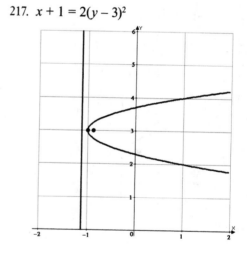

218. $y = -\frac{1}{8}(x-1)^2 + 4$

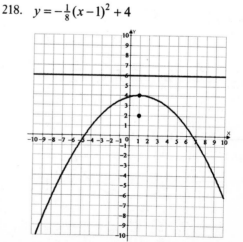

219. $x - 4 = \frac{1}{2}(y + 5)^2$

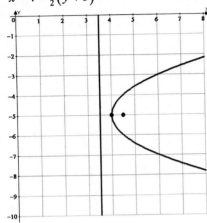

222. $4(x + 1)^2 + 4(y + 5)^2 = 100$

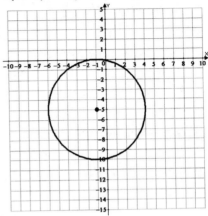

220. $x + 2 = -(y + 6)^2$

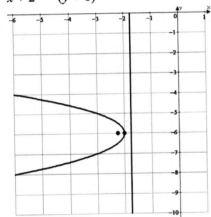

223. $(x - 7)^2 + (y + 1)^2 = 36$

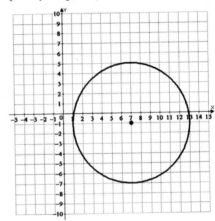

221. $(x - 8)^2 + (y + 8)^2 = 64$

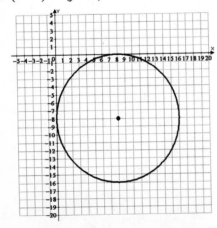

224. $x^2 + y^2 - 6y - 40 = 0$

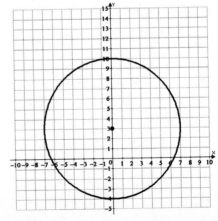

225. $x^2 + y^2 - 6x = 0$

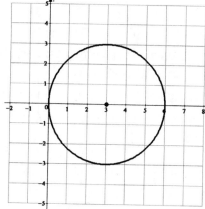

226. $4x^2 + 25y^2 = 100$

$$\frac{4x^2}{100} + \frac{25y^2}{100} = 1$$

$$\frac{x^2}{25} + \frac{y^2}{4} = 1$$

227. $16x^2 - 96x + y^2 + 128 = 0$

$$16\left(x^2 - 6x\right) + y^2 = -128$$

$$16\left(x^2 - 6x + 9\right) + y^2 = -128 + 144$$

$$16(x - 3)^2 + y^2 = 16$$

$$\frac{(x-3)^2}{1} + \frac{y^2}{16} = 1$$

228. $9x^2 + 4y^2 + 8y + 4 = 36$

$$9x^2 + 4\left(y^2 + 2y\right) = 32$$

$$9x^2 + 4\left(y^2 + 2y + 1\right) = 32 + 4$$

$$9x^2 + 4(y + 1)^2 = 36$$

$$\frac{x^2}{4} + \frac{(y+1)^2}{9} = 1$$

229. $9x^2 - 90x + 36y^2 + 144y + 45 = 0$

$$9\left(x^2 - 10x\right) + 36\left(y^2 + 4y\right) = -45$$

$$9\left(x^2 - 10x + 25\right) + 36\left(y^2 + 4y + 4\right) = -45 + 225 + 144$$

$$9(x - 5)^2 + 36(y + 2)^2 = 324$$

$$\frac{(x-5)^2}{36} + \frac{(y+2)^2}{9} = 1$$

230. $7x^2 + 4y^2 + 98x + 64y + 571 = 0$

$$7x^2 + 98x + 4y^2 + 64y = -571$$

$$7\left(x^2 + 14x\right) + 4\left(y^2 + 16y\right) = -571$$

$$7\left(x^2 + 14x + 49\right) + 4\left(y^2 + 16y + 64\right) = -571 + 7 \cdot 49 + 4 \cdot 64$$

$$7(x + 7)^2 + 4(y + 8)^2 = 28$$

$$\frac{(x+7)^2}{4} + \frac{(y+8)^2}{7} = 1$$

231. $\frac{(x+1)^2}{4} + \frac{(y-5)^2}{9} = 1$

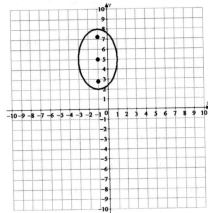

234. $\frac{(x-6)^2}{49} + \frac{(y+9)^2}{64} = 1$

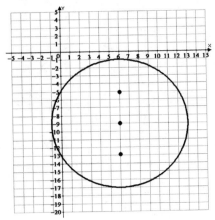

232. $\frac{(x-2)^2}{25} + \frac{(y-3)^2}{16} = 1$

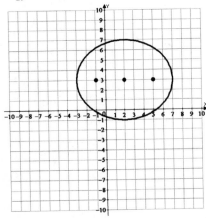

235. Given a center at (2, 1), focus at (2, 5), and vertex at (2, 6), there's a second focus at (2, –3) and a second vertex at (2, –4). In $\frac{(x-2)^2}{b^2} + \frac{(y-1)^2}{a^2} = 1$, $a = 5$, $c = 4$, and b can be found with $c^2 = a^2 - b^2$. $16 = 25 - b^2$, so $b = 3$, and the equation is $\frac{(x-2)^2}{9} + \frac{(y-1)^2}{25} = 1$.

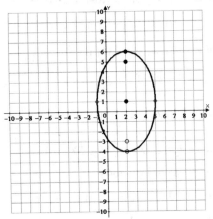

233. $\frac{(x+7)^2}{36} + \frac{(y+9)^2}{100} = 1$

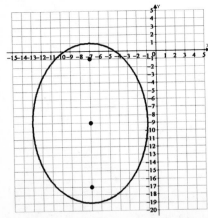

236. $x^2 - y^2 + 2x - 2y - 3 = 1$

$(x^2 + 2x) - (y^2 + 2y) = 4$

$(x^2 + 2x + 1) - (y^2 + 2y + 1) = 4 + 1 - 1$

$(x+1)^2 - (y+1)^2 = 4$

$\frac{(x+1)^2}{4} - \frac{(y+1)^2}{4} = 1$

237. $4x^2 - y^2 + 24x - 6y + 23 = 0$

$4(x^2 + 6x) - (y^2 + 6y) = -23$

$4(x^2 + 6x + 9) - (y^2 + 6y + 9) = -23 + 36 - 9$

$4(x + 3)^2 - (y + 3^2) = 4$

$\frac{(x+3)^2}{1} - \frac{(y+3^2)}{4} = 1$

238. $y^2 + 10y - 4x^2 + 56x = 187$

$(y^2 + 10y) - 4(x^2 - 14x) = 187$

$(y^2 + 10y + 25) - 4(x^2 - 14x + 49) = 187 + 25 - 4 \cdot 49$

$(y + 5)^2 - 4(x - 7)^2 = 16$

$\frac{(y+5)^2}{16} - \frac{(x-7)^2}{4} = 1$

239. $9x^2 - 16y^2 + 36x - 96y - 252 = 0$

$9(x^2 + 4x) - 16(y^2 + 6y) = 252$

$9(x^2 + 4x + 4) - 16(y^2 + 6y + 9) = 252 + 36 - 144$

$9(x + 2)^2 - 16(y + 3)^2 = 144$

$\frac{(x+2)^2}{16} - \frac{(y+3)^2}{9} = 1$

240. $25y^2 - 4x^2 + 100y + 56x - 196 = 0$

$25(y^2 + 4y) - 4(x^2 - 14x) = 196$

$25(y^2 + 4y + 4) - 4(x^2 - 14x + 49) = 196 + 25 \cdot 4 - 4 \cdot 49$

$25(y + 2)^2 - 4(x - 7)^2 = 100$

$\frac{(y+2)^2}{4} - \frac{(x-7)^2}{25} = 1$

241. $\frac{x^2}{25} - \frac{y^2}{64} = 1$

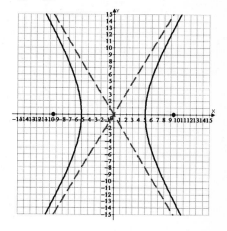

242. $\frac{(y+1)^2}{16} - \frac{(x-5)^2}{9} = 1$

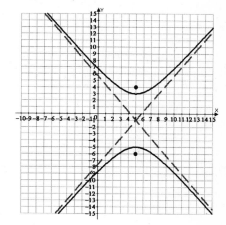

243. $\dfrac{(x+3)^2}{49} - \dfrac{y^2}{4} = 1$

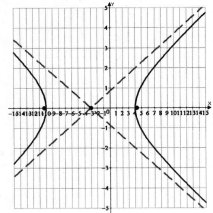

244. $81(y-6)^2 - 4(x+11)^2 = 324$

$$\dfrac{(y-6)^2}{4} - \dfrac{(x+11)^2}{81} = 1$$

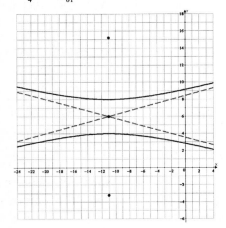

245. Given vertices at $(\pm 6, 2)$ and asymptotes $y = \pm\frac{5}{4}x + 2$, the center is $(0, 2)$, $a = 6$, $b = 7.5$, and $c \approx 9.6$. Equation is $\dfrac{x^2}{36} - \dfrac{(y-2)^2}{56.25} = 1$ or $\dfrac{x^2}{36} - \dfrac{4(y-2)^2}{225} = 1$.

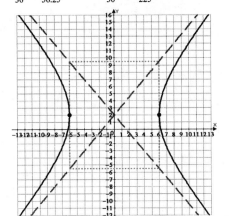

246. $4x^2 + 9y^2 = 36$

$y = -\frac{1}{2}x + 2$

247. $x^2 + y^2 = 16$

$9x^2 - 4y^2 = 36$

$\underline{9x^2 + 9y^2 = 144}$

$-13y^2 = -108$

$y^2 = \frac{108}{13}$

$y \approx \pm 2.88$

$x^2 + \frac{108}{13} = 16$

$x^2 = \frac{100}{13}$

$x \approx \pm 2.77$

248.
$$x^2 + y^2 = 4$$
$$x - 2y = 4 \rightarrow x = 4 + 2y$$
$$(4 + 2y)^2 + y^2 = 4$$
$$16 + 16y + 4y^2 + y^2 = 4$$
$$5y^2 + 16y + 12 = 0$$
$$(5y + 6)(y + 2) = 0$$
$$5y + 6 = 0 \qquad y + 2 = 0$$
$$y = -\tfrac{6}{5} \qquad y = -2$$

249.
$$y = x^2 - 4 \rightarrow x^2 = y + 4$$
$$x^2 + y^2 = 4$$
$$y + 4 + y^2 = 4$$
$$y^2 + y = 0$$
$$y(y + 1) = 0$$
$$y = 0 \qquad y = -1$$
$$x = \pm 2 \qquad x = \pm\sqrt{3}$$

250.
$$x + y = 2 \rightarrow x = 2 - y$$
$$x^2 - y^2 = 4$$
$$(2 - y)^2 - y^2 = 4$$
$$4 - 4y + y^2 - y^2 = 4$$
$$-4y = 0$$
$$y = 0$$
$$x = 2$$

Chapter 10

251. $64x^3 + 1 = (4x + 1)(16x^2 - 4x + 1)$

252. $8t^3 - 27 = (2t - 3)(4t^2 + 6t + 9)$

253. $8x^3 + 12x^2 + 6x + 1 =$
$(2x)^3 + 3(2x)^2(1) + 2(2x)(1^2) + 1^3 = (2x + 1)^3$

254. $27y^3 + 125 = (3y + 5)(9y^2 - 15y + 25)$

255. $x^3 - 343 = (x - 7)(x^2 + 7x + 49)$

256. Factors of constant: 1, 3. Factors of lead: 1, 2, 4. Possible rational zeros: $\pm 1, \pm 3, \pm\tfrac{1}{2}, \pm\tfrac{3}{2}, \pm\tfrac{1}{4}, \pm\tfrac{3}{4}$.

257. Factors of constant: 1, 2, 3, 4, 6, 12. Factors of lead: 1. Possible rational zeros: $\pm 1, \pm 2, \pm 3, \pm 4, \pm 6, \pm 12$.

258. Factors of constant: 1, 2, 3, 4, 6, 12. Factors of lead: 1, 2. Possible rational zeros: $\pm 1, \pm 2, \pm 3, \pm 4, \pm 6, \pm 12, \pm\tfrac{1}{2}, \pm\tfrac{3}{2}$.

259. Factors of constant: 1, 2, 4, 5, 10, 20. Factors of lead: 1. Possible rational zeros: $\pm 1, \pm 2, \pm 4, \pm 5, \pm 10, \pm 20$.

260. Factors of constant: 1, 2, 3, 6, 9, 18. Factors of lead: 1, 3. Possible rational zeros: $\pm 1, \pm 2, \pm 3, \pm 6, \pm 9, \pm 18, \pm\tfrac{1}{3}, \pm\tfrac{2}{3}$.

261. $f(x) = x^3 + 2x^2 - 25x - 50$, maximum of 1 positive real zero. $f(-x) = -x^3 + 2x^2 + 25x - 50$, maximum of 2 negative real zeros.

262. $g(x) = 4x^3 + 12x^2 - 9x - 27$, maximum of 1 positive real zero. $g(-x) = -4x^3 + 12x^2 + 9x - 27$, maximum of 2 negative real zeros.

263. $f(x) = x^3 + 4x^2 - 9x - 36$, maximum of 1 positive real zero. $f(-x) = -x^3 + 4x^2 + 9x - 36$, maximum of 2 negative real zeros.

264. $g(x) = 9x^3 + 18x^2 - 4x - 8$, maximum of 1 positive real zero. $g(-x) = -9x^3 + 18x^2 + 4x - 8$, maximum of 2 negative real zeros.

265. $f(x) = x^3 + 2x^2 + 9x + 18$, no positive real zeros. $f(x) = -x^3 + 2x^2 - 9x + 18$, maximum of 3 negative real zeros.

266.
$$\begin{array}{r} x+7 \\ x-3{\overline{\smash{\big)}\,x^2+4x-8}} \\ \underline{-\left(x^2-3x\right)} \\ 7x-8 \\ \underline{-(7x-21)} \\ 13 \end{array}$$

$x-3$ is not a factor.

267.
$$\begin{array}{r} 2x+5 \\ 3x-4{\overline{\smash{\big)}\,6x^2+7x-18}} \\ \underline{-\left(6x^2-8x\right)} \\ 15x-18 \\ \underline{-(15x-20)} \\ 2 \end{array}$$

$3x-4$ is not a factor.

268.
$$\begin{array}{r} 3x^2+x+4 \\ x-2{\overline{\smash{\big)}\,3x^3-5x^2+2x-1}} \\ \underline{-\left(3x^3-6x^2\right)} \\ x^2+2x \\ \underline{-\left(x^2-2x\right)} \\ 4x-1 \\ \underline{-(4x-8)} \\ 7 \end{array}$$

$x-2$ is not a factor.

269.
$$\begin{array}{r} x-3 \\ 2x^2-3x+2{\overline{\smash{\big)}\,2x^3-9x^2+11x-6}} \\ \underline{-\left(2x^3-3x^2+2x\right)} \\ -6x^2+9x-6 \\ \underline{-\left(-6x^2+9x-6\right)} \\ 0 \end{array}$$

$2x^2-3x+2$ is a factor. Denominator $= 0$.

270.
$$\begin{array}{r} a^3-a^2+2a-4 \\ a+2{\overline{\smash{\big)}\,a^4+a^3+0a^2+0a-1}} \\ \underline{-\left(a^4+2a^3\right)} \\ -a^3+0a^2 \\ \underline{-\left(-a^3-2a^2\right)} \\ 2a^2+0a \\ \underline{-\left(2a^2+4a\right)} \\ -4a-1 \\ \underline{-(-4a-8)} \\ 7 \end{array}$$

$a+2$ is not a factor.

271. $\underline{-2|1\ \ -5\ \ -6}$

   $\underline{\ \ \downarrow\ \ -2\ \ 14\ \ }$

   $1\ \ -7\ \ \underline{|8}$

$x=-2$ is not a zero.

272. $\underline{1|\ 3\ \ -4\ \ 1}$

   $\underline{\ \ \downarrow\ \ 3\ \ -1\ \ }$

   $3\ \ -1\ \ \underline{|0}$

$x=1$ is a zero.

273.
$$\begin{array}{r} 2\rfloor\ \ 1\ \ \ 2\ \ \ 3\ \ \ 4 \\ \downarrow\ \ 2\ \ \ 8\ \ \ 22 \\ \hline 1\ \ 4\ \ 11\ \ \lfloor 26 \end{array}$$
$x = 2$ is not a zero.

274.
$$\begin{array}{r} 3\rfloor\ \ 3\ \ \ -1\ \ \ 2\ \ \ \ 5 \\ \downarrow\ \ \ 9\ \ \ 24\ \ \ 78 \\ \hline 3\ \ \ 8\ \ \ 26\ \ \ \lfloor 83 \end{array}$$
$x = 3$ is not a zero.

275.
$$\begin{array}{r} 1\rfloor\ \ 2\ \ \ 0\ \ \ 1\ \ \ -3 \\ \downarrow\ \ 2\ \ \ 2\ \ \ 3 \\ \hline 2\ \ \ 2\ \ \ 3\ \ \ \lfloor 0 \end{array}$$
$x = 1$ is a zero.

276. $f(x) = 9x^4 + 4x^3 - 3x^2 + 2x$
$f(x) = x(9x^3 + 4x^2 - 3x + 2)$
$$\begin{array}{r} -1\rfloor\ \ 9\ \ \ 4\ \ \ -3\ \ \ 2 \\ \downarrow\ \ -9\ \ \ 5\ \ \ -2 \\ \hline 9\ \ \ -5\ \ \ 2\ \ \ \lfloor 0 \end{array}$$
$f(x) = x(x+1)(9x^2 - 5x + 2)$

Use the quadratic formula to solve $9x^2 - 5x + 2 = 0$. Zeros of $f(x)$ are $x = 0$, $x = -1$, and $x = \frac{5 \pm i\sqrt{47}}{18}$.

277. $g(x) = 15x^4 - 25x^3 + 10x^2$
$g(x) = 5x^2(3x^2 - 5x + 2)$
$g(x) = 5x^2(3x - 2)(x - 1)$
Zeros of $g(x)$ are $x = 0$ (multiplicity of 2), $x = 1$, and $x = \frac{2}{3}$.

278. $f(x) = 3x^3 - 13x^2 + 8x + 12$
$$\begin{array}{r} 2\rfloor\ \ 3\ \ \ -13\ \ \ 8\ \ \ 12 \\ \downarrow\ \ 6\ \ \ -14\ \ \ -12 \\ \hline 3\ \ \ -7\ \ \ -6\ \ \ \lfloor 0 \end{array}$$
$f(x) = (x-2)(3x^2 - 7x - 6)$
$f(x) = (x-2)(x-3)(3x+2)$
Zeros of $f(x)$ are $x = 2$, $x = 3$, and $x = -\frac{2}{3}$.

279. $g(x) = 3x^3 - 18x^2 + 33x - 18$
$$\begin{array}{r} 3\rfloor\ \ 3\ \ \ -18\ \ \ 33\ \ \ -18 \\ \downarrow\ \ 9\ \ \ -27\ \ \ 18 \\ \hline 3\ \ \ -9\ \ \ 6\ \ \ \lfloor 0 \end{array}$$
$g(x) = (x-3)(3x^2 - 9x + 6)$
$g(x) = 3(x-3)(x^2 - 3x + 2)$
$g(x) = 3(x-3)(x-2)(x-1)$
Zeros of $g(x)$ are $x = 3$, $x = 2$, and $x = 1$.

280. $f(x) = 2x^3 + 12x^2 + 22x + 12$
$f(x) = 2(x^3 + 6x^2 + 11x + 6)$
$$\begin{array}{r} -2\rfloor\ \ 1\ \ \ 6\ \ \ 11\ \ \ 6 \\ \downarrow\ \ -2\ \ \ -8\ \ \ -6 \\ \hline 1\ \ \ 4\ \ \ 3\ \ \ \lfloor 0 \end{array}$$
$f(x) = 2(x+2)(x^2 + 4x + 3)$
$f(x) = 2(x+2)(x+3)(x+1)$
Zeros of $f(x)$ are $x = -2$, $x = -3$, and $x = -1$.

281. $2x^3 + 17x^2 + 41x + 30 \le 0$

$$\begin{array}{r|rrrr} -5 & 2 & 17 & 41 & 30 \\ & \downarrow & -10 & -35 & -30 \\ \hline & 2 & 7 & 6 & \underline{0} \end{array}$$

$(x+5)(2x^2 + 7x + 6) \le 0$

$(x+5)(x+2)(2x+3) \le 0$

Interval	Test Point	$x + 5$	$x + 2$	$2x + 3$	$2x^3 + 17x^2 + 41x + 30$
$x < -5$	-6	$-$	$-$	$-$	$-$
$-5 < x < -2$	-3	$+$	$-$	$-$	$+$
$-2 < x < -1.5$	-1.7	$+$	$+$	$-$	$-$
$x > -1.5$	0	$+$	$+$	$+$	$+$

$2x^3 + 17x^2 + 41x + 30 \le 0$ when $x \le -5$ and when $-2 \le x \le -1.5$.

282. Using the same analysis in problem 281, $2x^3 + 17x^2 + 41x + 30 > 0$ when $-5 < x < -2$ and when $x > -1.5$.

283. $20x^4 + 13x^2 - 15 \le 0$

$(5x^2 - 3)(4x^2 + 5) \le 0$

$4x^2 + 5$ has no real zeros. The zeros of $5x^2 - 3$ are $x \approx \pm 0.77$.

Interval	Test Point	$5x^2 - 3$	$4x^2 + 5$	$20x^4 + 13x^2 - 15$
$x < -0.77$	-1	$+$	$+$	$+$
$-0.77 < x < 0.77$	0	$-$	$+$	$-$
$x > 0.77$	1	$+$	$+$	$+$

$20x^4 + 13x^2 - 15 \le 0$ when $-0.77 < x < 0.77$.

284. $x^3 + 5x^2 - 4x - 20 \geq 0$

     $x^2(x+5) - 4(x+5) \geq 0$

     $(x+5)(x^2-4) \geq 0$

     $(x+5)(x+2)(x-2) \geq 0$

Interval	Test Point	$x+5$	$x+2$	$x-2$	$x^3 + 5x^2 - 4x - 20$
$x < -5$	-6	$-$	$-$	$-$	$-$
$-5 < x < -2$	-3	$+$	$-$	$-$	$+$
$-2 < x < 2$	0	$+$	$+$	$-$	$-$
$x > 2$	3	$+$	$+$	$+$	$+$

$x^3 + 5x^2 - 4x - 20 \geq 0$ when $-5 \leq x \leq -2$ and when $x \geq 2$.

285. $45x^4 \leq 30x^3 - 5x^2$

     $45x^4 - 30x^3 + 5x^2 \leq 0$

     $5x^2(9x^2 - 6x + 1) \leq 0$

     $5x^2(3x-1)^2 \leq 0$

Interval	Test Point	$5x^2$	$(3x-1)^2$	$45x^4 - 30x^3 + 5x^2$
$x < 0$	-1	$+$	$+$	$+$
$0 < x < \frac{1}{3}$	0.1	$+$	$+$	$+$
$x > \frac{1}{3}$	1	$+$	$+$	$+$

$45x^4 \leq 30x^3 + 5x^2$ is equivalent to $45x^4 - 30x^3 + 5x^2 \leq 0$, and $45x^4 - 30x^3 + 5x^2$ is never less than 0.

286. $y = 2x^3 + 20x^2 + 50x$

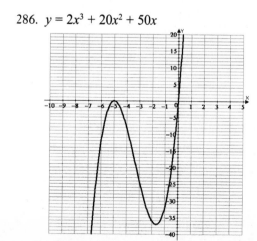

287. $y = x^3 + 5x^2 - 9x - 45$

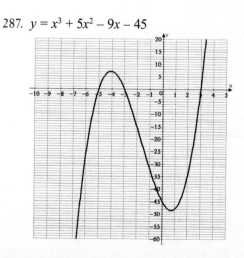

288. $y = x^3 - 17x^2 + 91x - 147$

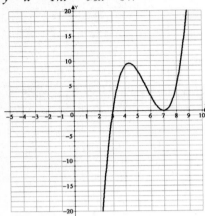

289. $y = 3x^3 + 2x^2 - 27x - 18$

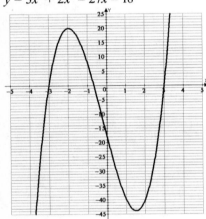

290. $y = x^3 + 5x^2 - 4x - 20$

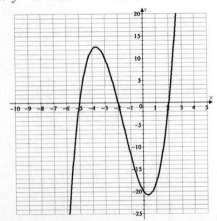

Chapter 11

291. Domain: $x \neq 5$, $\frac{t-5}{5-t} = \frac{\cancel{t-5}}{-1\cancel{(t-5)}} = -1$

292. Domain: $x \neq 5$, $\frac{x^2-25}{5-x} = \frac{(x+5)\cancel{(x-5)}}{-1\cancel{(x-5)}} = -(x+5)$

293. Domain: $x \neq \pm 2$, $\frac{y^2-y-6}{y^2-4} = \frac{(y-3)\cancel{(y+2)}}{(y-2)\cancel{(y+2)}} = \frac{y-3}{y-2}$

294. Domain: $x \neq -4$, $\frac{x^2-16}{6x+24} = \frac{\cancel{(x+4)}(x-4)}{6\cancel{(x+4)}} = \frac{x-4}{6}$

295. Domain: $x \neq -2, x \neq -6$, $\frac{x^2-4x-12}{x^2+8x+12} = \frac{(x-6)\cancel{(x+2)}}{(x+6)\cancel{(x+2)}} = \frac{x-6}{x+6}$

296. $\frac{-2}{x^2-2x-3} + \frac{3}{x^2-9} = \frac{-2(x+3)}{(x-3)(x+1)(x+3)} + \frac{3(x+1)}{(x+3)(x-3)(x+1)} = \frac{-2x-6+3x+3}{(x+3)(x-3)(x+1)}$

$= \frac{\cancel{x-3}}{(x+3)\cancel{(x-3)}(x+1)} = \frac{1}{(x+3)(x+1)} = \frac{1}{x^2+4x+3}$

297. $\frac{x-4}{2x-6} + \frac{3}{x^2-9} = \frac{(x-4)(x+3)}{2(x-3)(x+3)} + \frac{3\cdot 2}{(x+3)(x-3)\cdot 2} = \frac{x^2-x-12+6}{2(x-3)(x+3)} = \frac{x^2-x-6}{2x^2-18}$

298. $\dfrac{2x-4}{x^2+5x+4} - \dfrac{x-4}{x^2+6x+8} = \dfrac{2(x-2)(x+2)}{(x+1)(x+4)(x+2)} - \dfrac{(x-4)(x+1)}{(x+2)(x+4)(x+1)} = \dfrac{2x^2-8}{(x+1)(x+4)(x+2)} - \dfrac{x^2-3x-4}{(x+2)(x+4)(x+1)}$

$$= \dfrac{x^2+3x-4}{(x+1)(x+4)(x+2)} = \dfrac{\cancel{(x+4)}(x-1)}{(x+1)\cancel{(x+4)}(x+2)} = \dfrac{x-1}{x^2+3x+2}$$

299. $\dfrac{2x-2}{x^2+4x+3} - \dfrac{x-1}{x^2+5x+6} = \dfrac{2(x-1)(x+2)}{(x+3)(x+1)(x+2)} - \dfrac{(x-1)(x+1)}{(x+3)(x+2)(x+1)} = \dfrac{2x^2-2x-4}{(x+3)(x+1)(x+2)} - \dfrac{x^2-1}{(x+3)(x+2)(x+1)}$

$$= \dfrac{x^2-2x-3}{(x+3)(x+2)(x+1)} = \dfrac{\cancel{(x+1)}(x-3)}{(x+3)(x+2)\cancel{(x+1)}} = \dfrac{x-3}{x^2+5x+6}$$

300. $2 - \dfrac{9}{3x+1} = \dfrac{2}{1}\cdot\dfrac{3x+1}{3x+1} - \dfrac{9}{3x+1} = \dfrac{6x+2-9}{3x+1} = \dfrac{6x-7}{3x+1}$

301. $\dfrac{x+5}{x^2-25}\cdot\dfrac{x-5}{x^2-10x+25} = \dfrac{\cancel{x+5}}{\cancel{(x+5)}(x-5)}\cdot\dfrac{\cancel{x-5}}{(x-5)\cancel{(x-5)}} = \dfrac{1}{x^2-10x+25}$

302. $\dfrac{3y^2-3y}{3y-12}\cdot\dfrac{y^2-2y-8}{y^2+3y+2} = \dfrac{\cancel{3}y(y-1)}{\cancel{3}\cancel{(y-4)}}\cdot\dfrac{\cancel{(y-4)}(y+2)}{\cancel{(y+2)}(y+1)} = \dfrac{y^2-y}{y+1}$

303. $\dfrac{2x^2-4x}{2x^2-2}\cdot\dfrac{x^2-2x-3}{x^2-5x+6} = \dfrac{\cancel{2}x\cancel{(x-2)}}{\cancel{2}\cancel{(x+1)}(x-1)}\cdot\dfrac{\cancel{(x-3)}\cancel{(x+1)}}{\cancel{(x-2)}\cancel{(x-3)}} = \dfrac{x}{x-1}$

304. $\dfrac{3x-12}{x^2-4}\cdot\dfrac{x^2+6x+8}{x-4} = \dfrac{3\cancel{(x-4)}}{\cancel{(x+2)}(x-2)}\cdot\dfrac{(x+4)\cancel{(x+2)}}{\cancel{x-4}} = \dfrac{3x+12}{x-2}$

305. $\dfrac{y-1}{y^2-y-6}\cdot\dfrac{y^2+5y+6}{y^2-1} = \dfrac{\cancel{y-1}}{(y-3)\cancel{(y+2)}}\cdot\dfrac{(y+3)\cancel{(y+2)}}{(y+1)\cancel{(y-1)}} = \dfrac{y+3}{y^2-2y-3}$

306. $\dfrac{a^2-5a+6}{a^2-2a-3}\div\dfrac{a-5}{a^2+3a+2} = \dfrac{(a-2)\cancel{(a-3)}}{\cancel{(a-3)}\cancel{(a+1)}}\cdot\dfrac{\cancel{(a+1)}(a+2)}{a-5} = \dfrac{a^2-4}{a-5}$

307. $\dfrac{2x^2-5x-12}{4x^2+8x+3}\div\dfrac{x^2-16}{2x^2+7x+3} = \dfrac{\cancel{(2x+3)}\cancel{(x-4)}}{\cancel{(2x+1)}\cancel{(2x+3)}}\cdot\dfrac{\cancel{(2x+1)}(x+3)}{(x+4)\cancel{(x-4)}} = \dfrac{x+3}{x+4}$

308. $\dfrac{9t^2-1}{6t^2+7t-3}\div\dfrac{27t^3+1}{8t^3+27} = \dfrac{\cancel{(3t+1)}\cancel{(3t-1)}}{\cancel{(2t+3)}\cancel{(3t-1)}}\cdot\dfrac{\cancel{(2t+3)}(4t^2-6t+9)}{\cancel{(3t+1)}(9t^2-3t+1)} = \dfrac{4t^2-6t+9}{9t^2-3t+1}$

309. $\dfrac{x^5-x^2}{5x^5-5x}\div\dfrac{10x^4-10x^2}{2x^4+2x^3+2x^2} = \dfrac{x^3(x+1)(x-1)}{5x(x^4-1)}\cdot\dfrac{2x^2(x^2+x+1)}{10x^2(x+1)(x-1)} = \dfrac{x^3\cancel{(x+1)}\cancel{(x-1)}}{5x(x^2+1)\cancel{(x+1)}\cancel{(x-1)}}\cdot\dfrac{2x^2(x^2+x+1)}{10x^2(x+1)(x-1)}$

$$= \dfrac{2x^3(x^2+x+1)}{50x(x^2+1)(x^2-1)} = \dfrac{x^2(x^2+x+1)}{25(x^4-1)}$$

310. $\dfrac{4y^2-y^3}{16-y^2} \div \dfrac{64+y^3}{16+8y+y^2} = \dfrac{y^2\,(\cancel{4-y})}{(\cancel{4+y})\,(\cancel{4-y})} \cdot \dfrac{(\cancel{4+y})\,(\cancel{4+y})}{(\cancel{4+y})(16+4y+y^2)} = \dfrac{y^2}{16+4y+y^2}$

311. $\dfrac{\frac{1+2}{x\;y}}{\frac{2+1}{x\;y}} \cdot \dfrac{xy}{xy} = \dfrac{y+2x}{2y+x}$

312. $\dfrac{\frac{x-5}{x^2-4}}{\frac{x^2-25}{x+2}} \cdot \dfrac{(x+2)(x-2)}{(x+2)(x-2)} = \dfrac{\cancel{x-5}}{(x+5)(\cancel{x-5})(x-2)} = \dfrac{1}{x^2+3x-10}$

313. $\dfrac{1-\frac{1}{x}}{\frac{1}{x}} \cdot \dfrac{x}{x} = \dfrac{x-1}{1} = x-1$

314. $\dfrac{1+\frac{1}{x-2}}{1-\frac{3}{x+2}} \cdot \dfrac{(x+2)(x-2)}{(x+2)(x-2)} = \dfrac{(x+2)(x-2)+(x+2)}{(x+2)(x-2)-3(x-2)} = \dfrac{x^2-4+x+2}{x^2-4-3x+6} = \dfrac{x^2+x-2}{x^2-3x+2} = \dfrac{(x+2)(\cancel{x-1})}{(x-2)(\cancel{x-1})} = \dfrac{x+2}{x-2}$

315. $\dfrac{\frac{2a}{3a^3-3}}{\frac{4a}{6a-6}} \cdot \dfrac{6(a+1)(a-1)}{6(a+1)(a-1)} = \dfrac{2a\cdot2}{4a(a+1)} = \dfrac{\cancel{4a}}{\cancel{4a}(a+1)} = \dfrac{1}{a+1}$

316. $\dfrac{2}{a+5} = \dfrac{1}{3} \rightarrow a+5=6 \rightarrow a=1$

317. $\dfrac{x+1}{x+2} = \dfrac{x-3}{x-4} \rightarrow x^2-3x-4 = x^2-x-6 \rightarrow -3x-4 = -x-6 \rightarrow -2x=-2 \rightarrow x=1$

318. $\dfrac{-2}{x-3} = \dfrac{x+3}{4} \rightarrow x^2-9=-8 \rightarrow x^2-1=0 \rightarrow x=\pm1$

319. $\dfrac{3}{a-2} = \dfrac{2}{a-5} \rightarrow 3a-15 = 2a-4 \rightarrow a=11$

320. $\dfrac{y-2}{y-5} = \dfrac{2}{y+5} \rightarrow y^2+3y-10 = 2y-10 \rightarrow y^2+y=0 \rightarrow y(y+1)=0 \rightarrow y=0, y=-1$

321. No solution. $2(x+1)\left(\dfrac{x}{x+1}-\dfrac{1}{2}\right) = \left(\dfrac{-1}{x+1}\right)\cdot2(x+1)$

$$2x-(x+1)=-2$$
$$x-1=-2$$
$$x=-1 \ \text{reject}$$

322. $4(x+3)(x-3)\left(\frac{x}{x^2-9}-\frac{1}{x+3}\right)=\left(\frac{1}{4x-12}\right)\cdot 4(x+3)(x-3)$

$\qquad 4x-4(x-3)=x+3$

$\qquad 4x-4x+12=x+3$

$\qquad\qquad\qquad x=9$

323.

$$\frac{5}{x^2-3x+2}-\frac{1}{x-2}=\frac{1}{3x-3}$$

$3(x-2)(x-1)\left[\frac{5}{(x-2)(x-1)}-\frac{1}{x-2}\right]=\left[\frac{1}{3(x-1)}\right]\cdot 3(x-2)(x-1)$

$\qquad\qquad 15-3(x-1)=x-2$

$\qquad\qquad 15-3x+3=x-2$

$\qquad\qquad\qquad\quad 20=2x$

$\qquad\qquad\qquad\quad x=10$

324.

$$\frac{2}{x^2-7x+12}-\frac{1}{x^2-9}=\frac{4}{x^2-x-12}$$

$(x+3)(x-3)(x-4)\left[\frac{2}{(x-3)(x-4)}-\frac{1}{(x+3)(x-3)}\right]=\left[\frac{4}{(x-4)(x+3)}\right](x+3)(x-3)(x-4)$

$\qquad\qquad 2(x+3)-1(x-4)=4(x-3)$

$\qquad\qquad\quad 2x+6-x+4=4x-12$

$\qquad\qquad\qquad\quad x+10=4x-12$

$\qquad\qquad\qquad\quad -3x=-22$

$\qquad\qquad\qquad\quad x=\frac{22}{3}$

325.

$$\frac{3}{y-4}-\frac{2}{y+1}=\frac{5}{y^2-3y-4}$$

$(y+1)(y-4)\left[\frac{3}{y-4}-\frac{2}{y+1}\right]=\left[\frac{5}{(y-4)(y+1)}\right](y+1)(y-4)$

$\qquad\qquad 3(y+1)-2(y-4)=5$

$\qquad\qquad\quad 3y+3-2y+8=5$

$\qquad\qquad\qquad\quad y+11=5$

$\qquad\qquad\qquad\quad y=-6$

326. Domain of $f(x)=\frac{-1}{x+2}-3$ is $x\neq-2$.
Vertical asymptote: $x=-2$.
Horizontal asymptote: $y=0$.

327. Domain of $g(x)=\frac{x-3}{x^2-1}$ is $x\neq\pm1$.
Vertical asymptotes: $x=-1$ and $x=1$.
Horizontal asymptote: $y=0$.

328. Domain of $f(x) = \frac{-2x}{x+5}$ is $x \neq -5$.
Vertical asymptote: $x = -5$.
Horizontal asymptote: $y = -2$.

329. Domain of $g(x) = \frac{x^2-x-6}{x+1}$ is $x \neq -1$.
Vertical asymptote: $x = -1$.
Oblique asymptote: $y = x - 2$.

330. Domain of $f(x) = \frac{x^2-9}{x-3}$ is $x \neq 3$.
Vertical asymptote: None. (Hole at [3, 6].)
Oblique asymptote: $y = x + 3$.

331. $f(x) = \frac{4}{x+2} - 3$

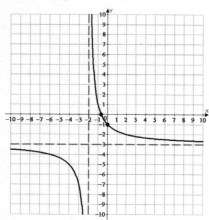

332. $g(x) = \frac{2x+1}{x-3}$

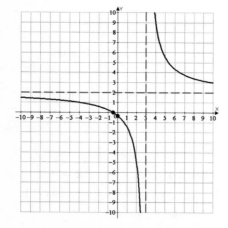

333. $f(x) = \frac{x+5}{x-3}$

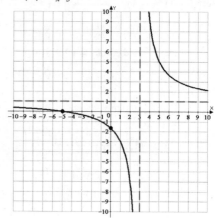

334. $g(x) = \frac{x^2-4}{x-3}$

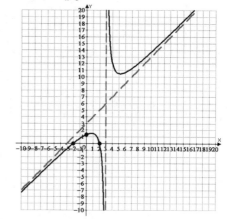

335. $f(x) = \frac{3x}{x^2-x-12}$

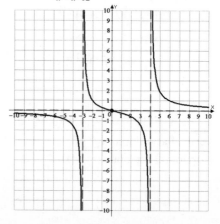

Chapter 12

336. $5^4 = 625 \rightarrow \log_5 625 = 4$

337. $9^y = 6,561 \rightarrow \log_9 6,561 = y$

338. $\log_2 1,024 = 10 \rightarrow 2^{10} = 1,024$

339. $\log_3 x = 2 \rightarrow 3^2 = x$

340. $\log_7 325 = \frac{\log 325}{\log 7} \approx 2.9723$

341. $\log_2\left(\frac{x^4}{y^3\sqrt{z}}\right) = 4\log_2 x - 3\log_2 y - \frac{1}{2}\log_2 z$

342. $\log\left(\frac{x^4 \cdot \sqrt[3]{y}}{z^3}\right) = 4\log x + \frac{1}{3}\log y - 3\log z$

343. $\log_8 \sqrt[3]{xy^5} = \frac{1}{3}\left(\log_8 x + 5\log_8 y\right) = \frac{1}{3}\log_8 x + \frac{5}{3}\log_8 y$

344. $\log_3\left(\frac{x^2 y}{\sqrt{z}}\right) = 2\log_3 x + \log_3 y - \frac{1}{2}\log_3 z$

345. $\log\left(\frac{x\sqrt{y}}{z^4}\right) = \log x + \frac{1}{2}\log y - 4\log z$

346. $3\log_4 x + \log_4 y - 2\log_4 z = \log_4\left(\frac{x^3 y}{z^2}\right)$

347. $2\log a + 3\log b - \frac{1}{3}\log c = \log\left(\frac{a^2 b^3}{\sqrt[3]{c}}\right)$

348. $\frac{1}{2}\log(x) + \log(y) - 4\log(z) = \log\left(\frac{\sqrt{x}y}{z^4}\right)$

349. $3\ln(x) + \ln(x+5) - 2\ln(x+4) = \ln\left(\frac{x^3(x+5)}{(x+4)^2}\right) = \ln\left(\frac{x^4+5x^2}{x^2+8x+16}\right)$

350. $2\log_2(x-1) + \log_2(x+5) - \log_2(x+3) = \log_2\left(\frac{(x-1)^2(x+5)}{x+3}\right)$

351. $8^{x+1} = 4$
$\left(2^3\right)^{x+1} = 2^2$
$2^{3x+3} = 2^2$
$3x+3 = 2$
$3x = -1$
$x = -\frac{1}{3}$

352. $9^{2x-4} = 27^3$
$\left(3^2\right)^{2x-4} = \left(3^3\right)^3$
$4x - 8 = 9$
$4x = 17$
$x = \frac{17}{4}$

353. $\log\left(12^{-x}\right) = \log(5)$
$-x\log 12 = \log 5$
$x = -\frac{\log 5}{\log 12} \approx -0.648$

354. $\log\left(3^{2x+1}\right) = \log(2)$
$(2x+1)\log 3 = \log 2$
$2x\log 3 + \log 3 = \log 2$
$2x\log 3 = \log 2 - \log 3$
$x = \frac{\log 2 - \log 3}{2\log 3} \approx -0.185$

355. $\log\left(10^{3x-4}\right) = \log\left(15^{x-2}\right)$
$(3x-4)\log 10 = (x-2)\log 15$
$3x - 4 = x\log 15 - 2\log 15$
$3x - x\log 15 = 4 - 2\log 15$
$x(3 - \log 15) = 4 - 2\log 15$
$x = \frac{4 - 2\log 15}{3 - \log 15} \approx 0.903$

356. $2^{2x} - 3\left(2^x\right) + 2 = 0$
$\left(2^x - 1\right)\left(2^x - 2\right) = 0$
$2^x - 1 = 0 \qquad 2^x - 2 = 0$
$2^x = 1 \qquad 2^x = 2$
$x = 0 \qquad x = 1$

357. $3\left(5^{2x}\right)+14\left(5^x\right)-5=0$

Let $a=5^x$

$3a^2+14a-5=0$

$(3a-1)(a+5)=0$

$3a-1=0$

$a=\frac{1}{3}$ $a+5=0$

$5^x=\frac{1}{3}$ $a=-5$

$x\log 5=\log\left(\frac{1}{3}\right)$ $5^x=-5$

$x=\frac{\log\left(\frac{1}{3}\right)}{\log 5}\approx-0.683$ *reject*

358. $2(3)^{2x}=\left(3^{x+1}\right)+20$

$2(3)^{2x}-\left(3^{x+1}\right)-20=0$

$2(3)^{2x}-3\left(3^x\right)-20=0$

Let $a=3^x$

$2a^2-3a-20=0$

$(2a+5)(a-4)=0$

$\qquad\qquad a-4=0$

$2a+5=0\qquad a=4$

$a=-\frac{5}{2}\qquad 3^x=4$

reject $x\log 3=\log 4$

$\qquad\quad x=\frac{\log 4}{\log 3}\approx1.262$

359. $e^{2x}-11\left(e^x\right)+30=0$

$\left(e^x-5\right)\left(e^x-6\right)=0$

$e^x-5=0\qquad e^x-6=0$

$e^x=5\qquad\quad e^x=6$

$x=\ln 5\approx1.609\qquad x=\ln 6\approx1.792$

360. $3\left(2^{2x}\right)+2^{x+1}-2^3=0$

$3\left(2^{2x}\right)+2\left(2^x\right)-8=0$

$\left(3\cdot2^x-4\right)\left(2^x+2\right)=0$

$3\cdot2^x-4=0$

$3\cdot2^x=4$ $2^x+2=0$

$2^x=\frac{4}{3}$ $2^x=-2$

$x\log 2=\log\left(\frac{4}{3}\right)$ *reject*

$x=\frac{\log\left(\frac{4}{3}\right)}{\log 2}\approx0.415$

361. $\log_2(x+2)+\log_2 x=3$

$\log_2\left(x^2+2x\right)=3$

$x^2+2x=2^3$

$x^2+2x-8=0$

$(x+4)(x-2)=0$

$x=-4$ $x=2$

reject

362. $\log_2(x+3)+\log_2 x=2$

$\log_2\left(x^2+3x\right)=2$

$x^2+3x=2^2$

$x^2+3x-4=0$

$(x+4)(x-1)=0$

$x=-4$ $x=1$

reject

363. $\log_3(x+3)-\log_3(x-1)=1$

$\log_3\left(\frac{x+3}{x-1}\right)=1$

$\frac{x+3}{x-1}=3^1$

$x+3=3x-3$

$6=2x$

$x=3$

364. $\log_8(x) + \log_8(x-3) = \frac{2}{3}$

$\log_8(x^2 - 3x) = \frac{2}{3}$

$x^2 - 3x = 8^{\frac{2}{3}}$

$x^2 - 3x = 4$

$x^2 - 3x - 4 = 0$

$(x-4)(x+1) = 0$

$x = 4 \qquad x = -1$

$\qquad\qquad$ *reject*

365. No solution. $\quad \log_4(x-2) - \log_4(x+1) = 1$

$\log_4\left(\frac{x-2}{x+1}\right) = 1$

$\frac{x-2}{x+1} = 4^1$

$4x + 4 = x - 2$

$3x = -6$

$x = -2$

reject

366. $f(x) = 2^{x-3} + 1$

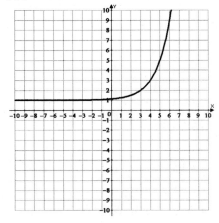

367. $g(x) = 4^{-x} - 3$

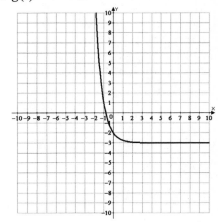

368. $f(x) = 3\log_2(x - 5)$

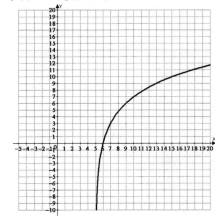

369. $g(x) = \log_3(2 - x) + 1$

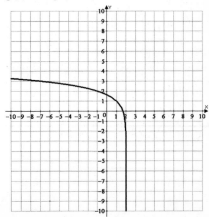

370. $f(x) = -2\ln(x + 3) - 4$

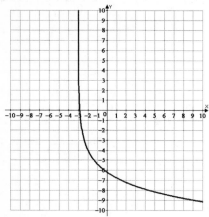

Log function shifted and inverted.

371. Invest about $10,605.91.

$$A = Pe^{rt}$$

$$31,231 = Pe^{0.06(18)}$$

$$\frac{31,231}{e^{0.06(18)}} = P$$

$$P \approx \$10,605.91$$

372. You will double your money in about 17 years and 4 months.

$$A = P\left(1 + \frac{r}{n}\right)^{nt}$$

$$2 = 1\left(1 + \frac{0.04}{12}\right)^{12t}$$

$$\log 2 = 12t \log\left(1 + \frac{0.04}{12}\right)$$

$$\frac{\log 2}{12\log(1.0033)} \approx t$$

$$t \approx 17.36 \text{ years}$$

373. After 7 days, there will be approximately 14,000 bacteria.

$$A = Pe^{rt}$$

$$900 = 300e^{r \cdot 2} \qquad A = Pe^{rt}$$

$$3 = e^{2r} \qquad A \approx 300e^{0.549 \cdot (7)}$$

$$\ln 3 = 2r \qquad A \approx 13,999.58$$

$$r = \frac{\ln 3}{2} \approx 0.549$$

374. A fossil that has 74% of its original Carbon 14 remaining is approximately 2,489 years old.

$$A = Pe^{rt}$$

$$\frac{1}{2} = 1e^{r \cdot 5,730}$$

$$\ln\left(\frac{1}{2}\right) = 5,730r$$

$$r = \frac{\ln\left(\frac{1}{2}\right)}{5,730} \approx -1.210 \times 10^{-4}$$

$$r \approx -0.0001210$$

$$A = Pe^{rt}$$

$$0.74 = 1e^{-0.0001210t}$$

$$\ln(0.74) = -0.0001210t$$

$$t = \frac{\ln(0.74)}{-0.0001210} \approx 2,489.128$$

375. The energy released in the 1906 earthquake was about 8 times $\left(\frac{6,309.573}{794.328} \approx 7.943\right)$ as large as that of the 1989 earthquake.

1906

$$M = \log\left(\frac{I}{I_0}\right)$$

$$7.8 = \log\left(\frac{I}{10^{-4}}\right)$$

$$7.8 = \log I - \log 10^{-4}$$

$$7.8 + \log 10^{-4} = \log I$$

$$7.8 - 4 = \log I$$

$$3.8 = \log I$$

$$I = 10^{3.8} \approx 6,309.573$$

1989

$$M = \log\left(\frac{I}{I_0}\right)$$

$$6.9 = \log\left(\frac{I}{10^{-4}}\right)$$

$$6.9 = \log I - \log 10^{-4}$$

$$6.9 + \log 10^{-4} = \log I$$

$$6.9 - 4 = \log I$$

$$2.9 = \log I$$

$$I = 10^{2.9} \approx 794.328$$

Chapter 13

376. $9 \cdot 10 \cdot 10 = 900$

377. $9 \cdot 10 \cdot 5 = 450$

378. $26 \cdot 10 \cdot 10 \cdot 10 \cdot 26 = 676,000$

379. $2 \cdot 3 \cdot 3 = 18$

380. Given four digits, there are $4 \cdot 3 \cdot 2 \cdot 1 = 24$ different four-digit prices. Given six digits, there are $6 \cdot 5 \cdot 4 \cdot 3 = 360$ different prices.

381. $_4P_2 = 12$

382. $_7P_7 = 7! = 5,040$

383. $_5C_2 = 10$

384. $_8C_8 = 1$

385. $_{22}C_5 = 26{,}334$

386. $_4C_3 = 4$

387. $_5C_3 = 10$

388. $_6C_2 = 15$

389. $_7C_3 = 35$

390. $_9C_6 = 84$

391. $P(\text{ace and ace}) = \frac{4}{52} \cdot \frac{4}{52} = \frac{1}{13} \cdot \frac{1}{13} = \frac{1}{169}$

392. $P(\text{ace and ace}) = \frac{4}{52} \cdot \frac{3}{51} = \frac{1}{13} \cdot \frac{1}{17} = \frac{1}{221}$

393. $P(\text{English and Spanish}) = \frac{125}{1{,}000} = \frac{5}{40} = \frac{1}{8}$

394. $P(\text{woman and only Spanish}) = \frac{150}{1{,}000} = \frac{3}{20}$

395. $P(\text{only English|Man}) =$

$$\frac{P(\text{Man and only English})}{P(\text{Man})} = \frac{210}{1{,}000} \div \frac{500}{1{,}000} = \frac{210}{1{,}000} \cdot \frac{1{,}000}{500} = \frac{210}{500} = \frac{21}{50}$$

396. $(x-2)^3 = 1x^3(-2)^0 + 3x^2(-2)^1 + 3x(-2)^2 + 1x^0(-2)^3$
$$= x^3 - 6x^2 + 12x - 8$$

397. $(2x+1)^3 = 1(2x)^3 \cdot 1^0 + 3(2x)^2 \cdot 1^1 + 3(2x) \cdot 1^2 + 1(2x)^0 \cdot 1^3$
$$= 8x^3 + 12x^2 + 6x + 1$$

398. $(2x+3)^4 = 1(2x)^4 \cdot 3^0 + 4(2x)^3 \cdot 3^1 + 6(2x)^2 \cdot 3^2 + 4(2x)^1 \cdot 3^3 + 1(2x)^0 \cdot 3^4$
$$= 16x^4 + 96x^3 + 108x^2 + 216x + 81$$

399. $(x+2)^5 = 1x^5 \cdot 2^0 + 5x^4 \cdot 2^1 + 10x^3 \cdot 2^2 + 10x^2 \cdot 2^3 + 5x \cdot 2^4 + 1x^0 \cdot 2^5$
$$= x^5 + 10x^4 + 40x^3 + 80x^2 + 80x + 32$$

400. $(2x-1)^6 = 1(2x)^6(-1)^0 + 6(2x)^5(-1)^1 + 15(2x)^4(-1)^2 + 20(2x)^3(-1)^3$
$$+ 15(2x)^2(-1)^4 + 6(2x)^1(-1)^5 + 1(2x)^0(-1)^6$$
$$= 64x^6 - 192x^5 + 240x^4 - 160x^3 + 60x^2 - 12x + 1$$

Index